Verified Signal Processing Algorithms
in MATLAB and C

Arie Dickman

Verified Signal Processing Algorithms in MATLAB and C

Advised by Israel Greiss

 Springer

Arie Dickman
Rishon Lezion, Israel

ISBN 978-3-030-93365-4   ISBN 978-3-030-93363-0   (eBook)
https://doi.org/10.1007/978-3-030-93363-0

This Springer imprint is published by the registered company Springer Nature Switzerland AG
The registered company address is: Gewerbestrasse 11, 6330 Cham, Switzerland

To Rina for her love and support, to Yonathan for the joy he brought into our lives and to Alon, Eran, and Michal

Preface

This book is for engineers in the field of signal processing who wish to implement in the shortest time working systems that are mostly built from a collection of building blocks, residing in an ASIC or FPGA firmware or in C language software, running on an SBC or DSP. The book provides the software path regarding implementation, keeping in mind design.

A substantial part of the material may be found in books, papers, or application notes, but to the best of my knowledge there is currently no such book that presents ready-to-use solutions to common signal processing problems under one binding.

Most of the graduates or advanced course students in engineering faculties join the industry; after graduating, they are less interested in theory or math, but in solving problems, they are the public to whom this book is directed.

After an extended experience with the industry, I know that not many engineers are familiar with both MATLAB and C programming, so this book may assist to either weakness.

After reading this book and practicing the enclosed examples, one will understand how to design using MATLAB and implement in C basic elements involved in signal processing and some in control design.

All MATLAB and C codes in the book are verified and almost all are original since processing times for all C codes are specified; the reader is able to estimate the processing time on his own target, by comparing it to the I5 2.9 GHz CPU used for measurement here.

The book does not pretend to present the optimal solution for every algorithm, but rather to offer simple and practical solutions, I believe in solutions that consume less than 50% of the hardware resources, so that when a new requirement appears, there is still no need to replace hardware.

This book is not directed to total beginners, and notions like FFT, IFFT, FIR, and IIR are assumed as known to the reader.

Rishon Lezion, Israel   Arie Dickman

Acknowledgment

I would like to thank Professor Anthony J. Weiss from Tel Aviv University, Israel, for his brilliant comments and useful critiques of this book.

I thank Mrs. Mary E. James from Springer for her continuous help and initiative which were essential for the publication of this book.

Thanks to MathWorks Inc. for letting me use the MATLAB software which is an integral part of this book.

Contents

Chapter 1
Overview of Signal Detection

Abstract In this chapter basic terms of noise and signal detection are explained, what to do in order to maximize the probability of detection and minimize the probability of false alarms, and how to improve the SNR for detection.

Keywords Signal · SNR · Noise · FFT · False alarm · Detection · Coherent · DDC · Probability

1.1 Introduction

In this chapter basic terms of noise and signal detection will be explained, what to do in order to maximize the probability of detection and minimize the probability of false alarms, and how to improve the SNR for detection.

1.2 About Signal Detection

When we need to detect signals using RF receivers, we have to consider the relationship between signals and noise, a well known expression for the minimum detectable signal at the receiver's input with respect to the thermal noise is

$$S = -174 + 10 * \log 10 (B) + NF + SNR \qquad (1.1)$$

where

S = sensitivity in dbm.
B = IF bandwidth before detection.
NF = receiver noise figure in db.
SNR = desired signal to noise power ratio in db of the desired signal.

Since noise is an inevitable addition to signals, it is desired to maximize the SNR ratio and detect the lowest possible signal power by our processing unit. After sampling a "wide" RF band that contains many channels, we should isolate a respective channel with the narrowest bandwidth that contains the minimal noise power, and the spectral content of the signal as well.

Good common practice suggests that the receiver gain is such that the thermal noise is larger than the quantization noise of the a2d by 6–10 db.

If we use narrow band processing, we will use a digital down converter (DDC) in order to achieve the above, and if wide band processing is used, we will use an appropriate shaping window followed by an FFT to achieve the above. The larger the FFT order, then less noise is contained in each bin, as the noise is divided between all bins that reside within the frequency range "seen" by the a2d.

When we detect signals, we should use a power threshold that will distinguish between signals and noise, since we do not want to detect noise as signal, let us check for a normal distribution noise what is the expected % or probability of false detections out of many trials versus the threshold used, a MATLAB code section to do this is.

Code 1.1

```
Fs = 100e6;
Fin = 25e6;
N = 8192;
t = (0:N-1)/Fs;
Res_fft = Fs/N;
Ind = Fin/Res_fft + 1;
Nt = 10000;
False = zeros(10,1);

for m = 1:10
 for k = 1:Nt

   noise = randn(N,1);        % normal distribution
%   noise = 2*rand(N,1) – 1;  % uniform distribution
   Fout = abs(fft(noise,N)).^2/N/N;
   Fout = Fout(1:N/2);
   Th = m*mean(Fout);
   if (Fout(Ind) > Th)
     False(m) = False(m) + 1;
   end

 end % for k
end   % for m

figure
plot(False/Nt,'r.-')
title ('False alarm probability versus threshold','fontweight','b')
xlabel ('X mean(noise bins)','fontweight','b')
```

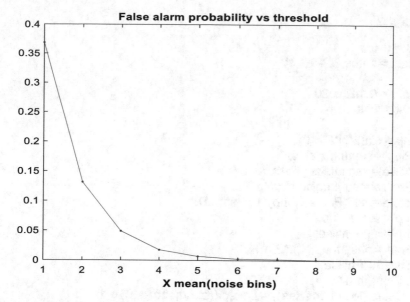

Fig. 1.1 False alarm probability versus signal to noise

The resulting figure shows that if a false alarm probability of less than 10^{-4} is desired (a common requirement), then a threshold of 10 times the average noise power per bin should be used, the same result is rendered if a uniform distribution for the noise is used (Fig. 1.1).

The next step is to determine the SNR ratio of a signal given the above threshold so that more than 99% of the trials will be successful, a MATLAB code section to do this is.

Code 1.2

```
Res = 0.1;
Detect = zeros(20/Res,1);

for snr = 0.1:Res:20
  for k = 1:Nt

    inp = sin(2*pi*Fin*t);
    Pinp = sum(inp.^2)/N;
    noise = randn(size(inp));
    Pnoise = sum(noise.^2)/N;
    scale = sqrt(Pnoise/Pinp)*10^(snr/20);
    inp = scale * inp;
    inp = inp + noise;
    Fout = abs(fft(inp,N)).^2/N/N;
    Th = 10*Pnoise;
    if (Fout(Ind) > Th)
      Detect(round(snr/Res)) = Detect(round(snr/Res)) + 1;
    end

  end
end

figure
plot(0.1:Res:20,Detect/Nt, 'r.-')
title ('Detect probability for 10 db threshold, Pfa < 0.0001 ','fontweight','b')
xlabel ('SNR','fontweight','b')
```

The resulting figure shows that for the above threshold and SNR > 13 db, more than 99% (a common requirement) of the trials are declared as "signal," the same result is rendered if a uniform distribution for the noise is used (Fig. 1.2).

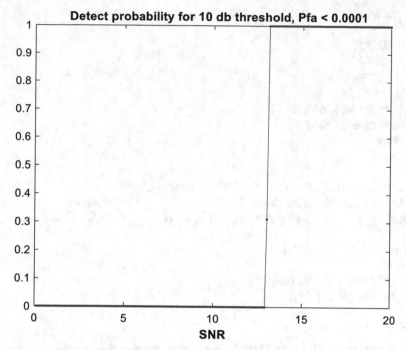

Fig. 1.2 Detection probability versus SNR for 10 db threshold

When the SNR of the signal is too low to be detected using a single snapshot, the time interval that we sample and observe should be increased, as known from the Cramer–Rao bound, estimation error for signals is inversely proportional to the square root of SNR and processing time.

If we use wide band processing by FFT, then we have to sum the FFT results frame by frame in order to increase the SNR, summing the respective complex result bin by bin is called coherent processing and summing the power result bin by bin is called non-coherent processing, a MATLAB code section that does both looks as.

Code 1.3

```
len = 4096;
Fs = 1.024e6;
Fin = 200*Fs/len;
t = (0:len-1)/Fs;
Fout1 = zeros(1,len);
Fout2 = zeros(1,len);
N = 10000;
for I = 1:N

  new = 0.1*cos(2*pi*Fin*t + rand(1)) +  randn(1,len);
  Fout1 = Fout1 + fft(new,len)/len;                    % Coherent
  Fout2 = Fout2 + abs(fft(new,len)).^2/len/len;        % Non Coherent

end

Fout1 = abs(Fout1(1:2048)).^2;
Fout2 = Fout2(1:2048);
```

The SNR for one frame is 12 db for both cases, the resulting figures show that for coherent processing the SNR is $10 * \log10(R)$ db better than for the non-coherent case where R is the ratio between the number of frames in both cases.

The noise power is increased by N for both cases, but the signal power is increased by N^2 for the coherent case and by N for the non-coherent case, so the SNR for the non-coherent case remains the same by summing frames, but for the coherent case it is increased by N (Figs. 1.3 and 1.4).

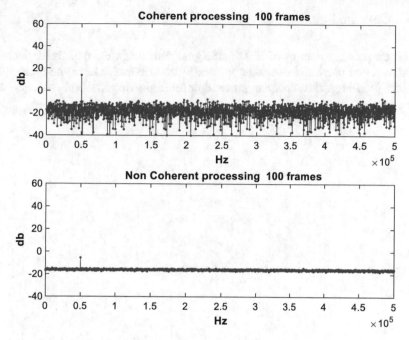

Fig. 1.3 Coherent and non-coherent processing for 100 frames

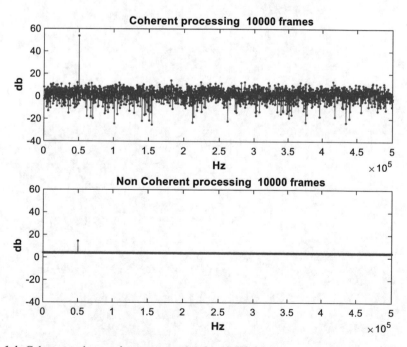

Fig. 1.4 Coherent and non-coherent processing for 10,000 frames

1.3   Conclusions

In this chapter basic terms of noise and signal detection were explained, decision thresholds that maximize detection probability and minimize false alarm probability were presented, then coherent processing for improving SNR for detection was demonstrated.

Chapter 2
Measures for Verification and Implementation of Algorithms

Abstract In this chapter measures to ensure that a designed algorithm meets its specifications with minimum resources are described and demonstrated.

Keywords Design · Verification · Frequency · Time · Coefficient file · Processing time · DDC

2.1 Introduction

In this chapter measures to ensure that a designed algorithm meets its specifications with minimum resources are described and demonstrated.

2.2 Measures Details

It is the responsibility of the designer to take measures to ensure that a designed algorithm meets its specifications with minimum resources, such as memory usage, transient time, and CPU processing time.

For demonstration, suppose that we designed two stages of decimation by two Firs, with final bandwidth of 30 Mhz, following a receiver with IF frequency of 120 Mhz sampled by a 160 Mhz a2d.

2.2.1 Algorithm Verification in Frequency

Insert a variable frequency sine wave to the algorithm, where the frequency sweeps around the IF frequency within a Nyquist zone and check the response, a MATLAB code section to do it is.

Code 2.1

```
Fs1 = 160e6;
Tpro = 1e-3;
N1 = round(Fs1*Tpro);
time_vec = (0:N1 – 1)/Fs1;
Fstart = -40e6;
Fstop = 40e6;
Res = 0.1e6;
offset = -Fstart/Res + 1;
IF = 120e6;
scx = IF + (Fstart:Res:Fstop);
ddc = exp(-j*2*pi*IF*time_vec);

for k = Fstart:Res:Fstop

inp1 = sin(2*pi*(k + IF)*time_vec + rand(1));
inp_pow(k/Res + offset) = sum(abs(inp1).^2)/length(inp1);
inp1 = inp1.*ddc;

y1 = filter(b13q,1,inp1);  % Fir1
y2 = y1(1:2:end);          % Down to 80 Mhz
y2 = filter(b23q,1,y2);    % Fir2
y6 = y2(1:2:end);          % Down to 40 Mhz
y6 = y6(40:end);           % Ignore transient

out_pow(k/Res + offset) = 2*sum(abs(y6).^2)/length(y6)/inp_pow(k/Res + offset);
% half of spectrum is on imag

end

figure, hold on, grid
plot (scx,10*log10(out_pow),'r.-')
axis ([IF+Fstart IF+Fstop -140 10])
title ('Freq response agter sweep','fontweight','bold')
xlabel('Hz','fontweight','bold')
ylabel('Channel response [db]','fontweight','bold')
hold off
```

Running that code renders the following figure (Fig. 2.1)

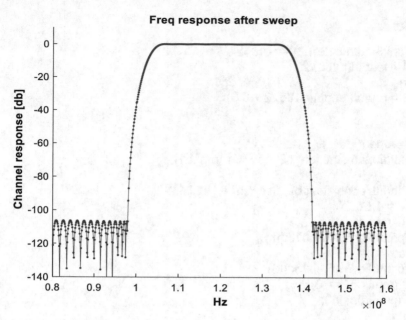

Fig. 2.1 Frequency response of a designed filter

2.2.2   *Preparing Coefficients File for C Software*

For C implementation, filter coefficients are saved in an *.h file, a MATLAB code section to do it is.

Code 2.2

```
if (rem(length(b13q),2) == 0)  % even
  L1 = length(b13q)/2;
else                            % odd
  L1 = round(length(b13q)/2 + 0.5);
end

fid = fopen('dec_fir_coeff.h','wt');
fprintf(fid,'#define len_b2_dec %d \n\n',L1);

fprintf(fid,'const float b2_dec[%d] = {\n',L1);
for k = 1:L1
  if (k == L1)
    fprintf(fid,'%fF\n',b13q(k));
  else
    fprintf(fid,'%fF,',b13q(k));
    if (rem(k,5) == 0)
      fprintf(fid,'\n');
    end
  end
end
fprintf(fid,'};\n\n');
fclose(fid);
```

2.2.3 Algorithm Verification in Time

After implementing the algorithm in C, correctness has to be verified, a signal which is suitable for the algorithm should be prepared (for this example a sine wave array) and converted to a binary file read by the software on the target on which the algorithm is performed.

Preparing the input file (in this case the type is float) in MATLAB looks like.

Code 2.3

```
time_vec = (0:9999)/Fs1;
IF = 8e6;
nco_out = 1000*sin(2*pi*IF*time_vec);

fid = fopen('nco_out.bin','wb');
fwrite(fid,nco_out,'float32');
fclose(fid);
```

Loading, operating the algorithm and saving to a binary file in C looks like.

Code 2.4

```
#include "stdio.h"      /* For files */
#include <stdlib.h>   /* For exit */
#include "dec_fir_coeff.h"
void alg(float *x, int n, float *y);  /* Prototype */
float inp_file[10000];
float out_file[10000];
int main()
{
int i,n;
FILE *fid;

fid = fopen("nco_out.bin","rb");
n = 0;
while (!feof(fid))
  fread(&inp_file[n++],4,1,fid);
fclose(fid);

alg((float *)inp_file, n, (float *)out_file);

fid = fopen("dec_Log.bin","wb");
fwrite(out_file, 4, n >>1, fid);
fclose(fid);
}
```

Now read the output file in MATLAB and compare to the MATLAB result as follows:

Code 2.5

```
fid = fopen('dec_Log.bin','rb');
y_soft = fread(fid,inf,'float32');
fclose(fid);

y_mat = filter(b13,1,nco_out);  % Fir1
y_mat = y_mat(1:2:end);         % Down to 80Mhz
y_mat = filter(b23,1,y_mat);    % Fir2
y_mat = y_mat(1:2:end);         % Down to 40Mhz
```

When comparing graphically on the same plot the arrays y_soft and y_mat, the software latency should be considered.

2.2.4 *Algorithm Transient Time Measurement*

The step response of the algorithm in terms of output samples is a good measure to the transient time, a MATLAB code section to do it is.

Code 2.6

```
N = 500;
x = 2*ones(1,N);
y1 = filter(b13q,1,x);   % Fir1
y2 = y1(1:2:end);        % Down to 80Mhz
y2 = filter(b23q,1,y2); % Fir2
y6 = y2(1:2:end);        % Down to 40Mhz
```

Running that code renders the following figure, which shows a transient response of about 20 output samples (Fig. 2.2).

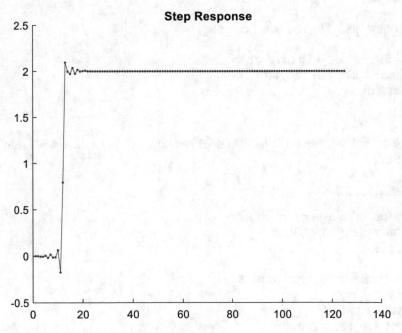

Fig. 2.2 Step response of a designed filter

2.2.5 Algorithm Processing Time Measurement

The goal of the designer and programmer is to develop an algorithm that will run in minimum time and use minimum memory resources, therefore enable the use of a cheaper CPU or ability to compute several channels in parallel.

The processing time should be measured on the target on which the algorithm is implemented.

A respective code section in C to do that is.

Code 2.7

```
#include "stdio.h"   /* For files */
#include <time.h>   /* For elapsed time calc */
clock_t  start,finish;
double duration;
int main()
{
int  i;
start = clock();

for (i = 0; i < 1000; i++)
  {
  alg((float *)inp_file, n, (float *)out_file);
  }

finish = clock();
duration = (double)(finish – start) / CLOCKS_PER_SEC;
}
```

Since the resolution of the clock_t measurement is mS, the algorithm may have to run 1000 or more sequential times to get the result in µS or nS for one run.

2.3   Conclusions

In this chapter measures to ensure that a designed algorithm meets its specifications were presented, the necessity of each measure depends on the algorithm.

Chapter 3
Narrow Band & Wide Band Processing Basics

Abstract In this chapter we describe and implement the narrow band processing elements NCO and FIR decimators, and how to easily find the image frequency that should be rejected. In this chapter we describe and implement the wide band processing elements of various shaping windows and FFT implementation types.

Keywords NCO · DDC · FIR · Decimation · CIC · Compensator · Aliasing · Image · Nyquist · Convolution · Shaping window · wola · FFT · bro · dro · Twiddle factor · Radix 2 · Radix 4

RF frequencies in some certain band or preselector are usually converted by a receiver to a fixed IF frequency followed by an analog filter at the desired bandwidth, higher bandwidth enables more frequencies in parallel but is more sensitive with respect to dynamic range and spurs.

Using digital signal processing, the IF analog information is converted to bits by an a2d converter, sampling at Fs Hz. Usually the IF frequencies are located at Fs/4, 3*Fs/4, 5*Fs/4... called Nyquist zones so that a maximum and symmetric bandwidth is attained.

Higher IF helps the receiver designer to avoid spurs but needs a more expensive a2d that has acceptable SNR and ENOB, as the ratio between Fs and IF is lower, jitter increases and decreases SNR at the a2d output.

There are two basic ways to process the IF frequency band, one is the narrow band processing, described by the following diagram:

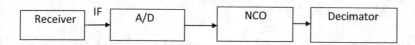

The digital IF samples stream is shifted to baseband, where it is filtered and decimated to the desired output bandwidth and sampling frequency, each channel requires dedicated processing.

The second way is to use wide band processing as described by the following diagram:

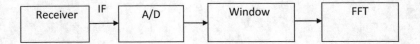

The digital IF sample stream undergoes a frequency shaping window that renders separation of channels (bins), then an N size FFT is performed which results in one complex number per bin.

If a stream of such complex numbers per bin over time is desired, then consecutive FFTs shifted in time are performed.

3.1 Narrow Band Processing

3.1.1 Introduction

In this chapter we describe and demonstrate design considerations and implementation of numerically controlled oscillator (NCO) and decimators which are the elements of narrow band processing, and how to easily find the image frequency that should be rejected when shifting frequencies.

3.1.2 Numerically Controlled Oscillator

The numerically controlled oscillator (NCO) shifts digitally the digital stream from the a2d by f1, making the operation in MATLAB as follows:

```
j = sqrt(-1);
t = (0:N-1)/Fs;
Nco_out = a2d_out.*exp(-j*2*pi*f1*t);
```

This operation generates an image frequency in addition to baseband, this image should be rejected by the first LPF (low pass filter) after the NCO, since the passband and stopband frequencies are also determined by the decimation rules then the strict requirement determines.

The generated image frequency is equal to twice f1 and aliased to (3*Fs – 2*f1), in order to not bother with calculating the resulting frequencies, a simple MATLAB code renders the correct image frequency as follows:

Code 3.1.1

```
Fs = 160e6;
f12 = 220e6;
f11 = f12 – 2e6;
f13 = f12 + 2e6;
N = 8192;
t = (0:N-1)/Fs;
inp = sin(2*pi*f11*t) + sin(2*pi*f12*t) + sin(2*pi*f13*t);
nco = inp.*exp(-j*2*pi*f12*t);
win = chebwin(N)';
win = win/sum(win);
Sigw = nco.*win;
Fout = abs(fftshift(fft(Sigw,N))).^2/N/N;
Fout = Fout/max(Fout);
Fx = (-4095:4096)*Fs/N;
```

That renders the following figure:

As seen from Fig. 3.1, the image frequency aliased to (3 * 160–2 * 220) Mhz should be rejected.

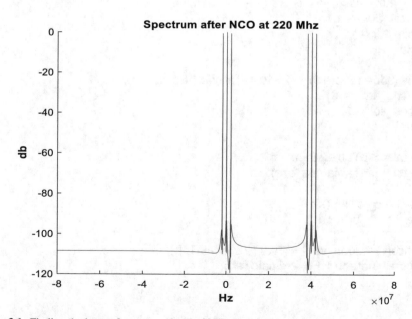

Fig. 3.1 Finding the image frequency after the NCO operation

For implementing the NCO operation, a sin table from 0 to 2*pi with 2^ScaleTab entries is used, described by a MATLAB code as

```
ScaleTab = 12;
SinTable = round(sin([0:2^ScaleTab -1)]/(2^ScaleTab)*2*pi)*(2^ScaleTab – 1));
```

The table coefficients are written to an *.h file as described in Sect. 2.2.2 before, ScaleTab is chosen by the required resolution of the frequency shifting, which is the sampling frequency of the a2d divided by the table size 2^ScaleTab.

Another consideration to increase ScaleTab is that the spectrum of the complex NCO output will not have spurs or elevated noise floor.
A MATLAB code that implements NCO operation is

Code 3.1.2

```
acc = 0;
IF = Fs/8;
Fc = round(IF/Fs*2^ScaleTab);

for k = 1:length(Inp)

  sin_adr = acc;
  if (sin_adr == 0)
   sin_adr = 1;
  end

  cos_adr = mod(acc + (2^ScaleTab)/4,2^ScaleTab);
  if (cos_adr == 0)
   cos_adr = 1;
  end

  s_tbl = SinTable(sin_adr);
  c_tbl = SinTable(cos_adr);

  var1 = c_tbl*Inp(k);
  var2 = -s_tbl*Inp(k);

  Xnco(k) = floor((var1 + j*var1)/2^ScaleTab);
  acc = mod(acc + Fc, 2^ScaleTab);

end
```

The cos() value is found from the sin table using the relation cos(a) = sin(pi/2 + a).
Suppose that a 16 bit a2d is used, the sin table we used has 2^15 elements of type short, a C code that implements the NCO operation is

Code 3.1.3

```
void nco(short *x, int n, int dp, short *y)
{
int sin_val, cos_val, acc, i;

acc = 0;
for (i = 0; i < n; i++)
  {
  sin_val = Sin_Tab[acc];
  cos_val = Sin_Tab[(acc + 8192) & 0x7fff];
  y[i] = (cos_val*x[i]) >> 15;
  y[I + n] = (-sin_val*x[i]) >> 15;
  acc = (acc + dp) & 0x7fff;
  }
}
```

The phase increment dp is round(f1/Fs*2^ScaleTab), the software produces a short array of the real part followed by a short array of the imag part.

If f1 is negative, then dp is round((Fs + f1)/Fs*2^ScaleTab), the above verified code consumes 31 nS per sample on an I5 2.9 GHz.

3.1.3   Finite Impulse Response Decimators

A decimator reduces the input sampling frequency to a lower output one, this operation requires a low pass filter (LPF) at the input sampling rate, followed by a suitable decimation. The highest passband frequency Fp at the LPF output should not exceed half of the output sampling frequency if the decimator is real, and not exceed the output sampling frequency if the decimator is complex, as per Nyquist theorem.

As we will see later, the most efficient way to design a decimator in terms of processing time is to divide the decimation between several stages or multirate method, the first stage should follow the rule of thumb

Fp = min(decimation constraint as above, image frequency of the NCO)

$$Fstop = Fp + 2^* \left(Fs / \left(2^* dec1 \right) - Fp \right)$$

where dec1 is the 1st stage decimation, for the next stages if any, the same rule applies besides rejecting the image frequency of the NCO.

Decimators are usually implemented by FIRs with possible addition of cascaded integrator comb (CIC) filters, they may also be implemented by IIRs, all types will be described later.

Earlier versions of the MATLAB used functions such as Remez, Firls, the current one uses Firpm to design filters, we found that using the fdesign tool in the signal processing toolbox renders efficient designs in short time, a MATLAB code that does it is

Code 3.1.4

```
Fs1 = 160e6;
dm = fdesign.lowpass('Fp,Fst,Ap,Ast',20e6,79e6,0.1,105,Fs1);
hm = design(dm);
b13 = hm.numerator;
b13 = b13/(sum(b13) + eps);
```

Fp is the largest passband frequency at which the filter gain is -1 db, lower value relative to the sampling frequency Fs1 renders more coefficients for the designed FIR.

Fstop is the smallest stopband frequency, the ratio between Fstop and Fp is called the shape factor of the filter where a value of less than 1.5 for the total response is common, reducing shape factor renders more coefficients for the designed FIR, Fstop should be less than Fs1/2.

Ap is the ripple in db at the passband, a value of 0.1 is common, reducing Ap renders more coefficients for the designed FIR.

Ast is the attenuation in db at the stopband, values between 90 and 110 are common to get actually -80 to -100 db, lower attenuation renders less coefficients for the designed FIR.

3.1.4 *Low Decimation Ratio Finite Impulse Response Filters*

When we say low decimation we mean 2–8.

We will demonstrate the design and verification by an example: Suppose that we need to design a decimator by 4, with final bandwidth of 30 Mhz, following a receiver with IF frequency of 120 Mhz sampled by a 160 Mhz a2d.

The efficient design in terms of processing time is to use multirate design, which is to divide the decimation as close as possible to decimation by 2 units, for our example that means 2 FIRs, the 1st closer to the a2d or NCO receives a double rate of samples relative to the second, but its shape factor requirement is lower, the second and final FIR has a shape factor of 1.47 measured from the total response.

A CIC filter, declared by the Z transform response $1/M * (1 - Z^{-M})/(1 - Z^{-1})$ where M is the filter order may be used instead of the 1st FIR, but the droop of this filter even for $M = 2$ is such that the total response cannot have a passband ripple of 0.1 db.

Trial and error method should be used by watching the figures of total frequency response and the frequency verification measure in Sect. 2.2.1, as seen from the total response the decimation rules of thumb should be kept at the end points Fp and Fstop but may be violated elsewhere.

The two FIRs design looks like

<u>Code 3.1.5</u>

```
res = 2048;
Fs1 = 160e6;
dec1 = 2;

F1 = Fs1/2/dec1/res*(0:res*dec1 − 1);
dm = fdesign.lowpass('Fp,Fst,Ap,Ast',20e6,79e6,0.1,105,Fs1);
hm = design(dm);
b13 =  hm.numerator;
b13 = b13/(sum(b13) + eps);
ha1 = freqz(b13,1,F1,Fs1);
```

That renders 9 coefficients and

```
F2 = Fs1/2/dec1/res*(0 : (res − 1);
dm = fdesign.lowpass('Fp,Fst,Ap,Ast',13.87e6,22e6,0.1,105,Fs1/2);
hm = design(dm);
b23 =  hm.numerator;
b23 = b23/(sum(b23) + eps);
ha2 = freqz(b23q,1,F2,Fs1/2);
```

That renders 42 coefficients

```
hat = ha1(1:length(F2)).*ha2;
```

The second filter is computed with the same resolution as the 1st but on half of frequencies, the total response is calculated on the same points and plotted on the second range as follows (Fig. 3.2).

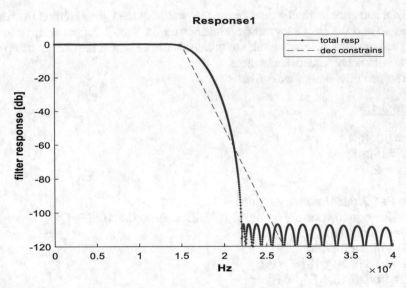

Fig. 3.2 Total frequency response for floating point FIR implementation

After design verification in frequency as described in Sect. 2.2.1 a C implementation for floating point inputs and coefficients which differs for even and odd number of coefficients looks like

Code 3.1.6

```
void dec_even(float *x, int n, int dec, float *y)
{
int i, k, count, len2;
float sum1,sum2;

count = 0;
len2 = 2*len_b_dec;
for (k = (len2 − 1); k < n; k += (2*dec))
  {
  sum1 = (float)0.;
  sum2 = (float)0.;

  for (i = 0; i < len_b_dec; i++)
    {
    sum1 += b_dec[i] * (x[k − i] + x[k − len2 + i + 1]);
    sum2 += b_dec[i] * (x[k − i + dec] + x[k − len2 + i + 1 + dec]);
    }

  y[count++] = sum1;
  y[count++] = sum2;
  }
}
void dec_odd(float *x, int n, int dec, float *y)
{
int i, k, count, len2;
float sum1,sum2;

count = 0;
len2 = 2*len_b_dec − 1;
for (k = (len2 − 1); k < n; k += (2*dec))
  {
  sum1 = (float)0.;
  sum2 = (float)0.;

  for (i = 0; i < (len_b_dec − 1); i++)
    {
    sum1 += b_dec[i] * (x[k − i] + x[k − len2 + i + 1]);
    sum2 += b_dec[i] * (x[k − i + dec] + x[k − len2 + i + 1 + dec]);
    }

  sum1 += b_dec[len_b_dec − 1]*x[k − len_b_dec + 1];
  sum2 += b_dec[len_b_dec − 1]*x[k − len_b_dec + 1 + dec];

  y[count++] = sum1;
  y[count++] = sum2;
  }
}
```

Two outputs are performed for each loop step, so the number of steps is half, if several computing units are available as in some DSPs, then more outputs could be computed every step.

Len_b_dec is half the filter length for even and half the filter length + 1 for odd length numbers.

The processing time is 74 µS for the 1st filter and 160 µS for the 2nd filter, both for a sampling interval of 60 µS (10,000 samples) on an I5 2.9G CPU, since the processing time is about 4 times larger than the sampling interval, one has to use a 4 times faster CPU to enable full throughput on a single core or use hardware implementation. Processing at 160 Mhz is too fast for current CPU technology.

If the processing unit does not have an efficient floating point support, then the computations may be done in fixed point format.

Representing the coefficients in fixed point format may be done in MATLAB by

$$b13 = \text{round}\left(b13^{*}\left(2^{\wedge}qu - 1\right)\right) / \left(2^{\wedge}qu - 1\right)$$

where qu is the number of bits used, when the number of coefficients increases some of the coefficients are very small which requires more bits, the same applies to stopband attenuation, where larger one requires more bits.

In the next figure the response for floating point coefficients and 16 bits fixed point coefficients for the above example are presented (Fig. 3.3).

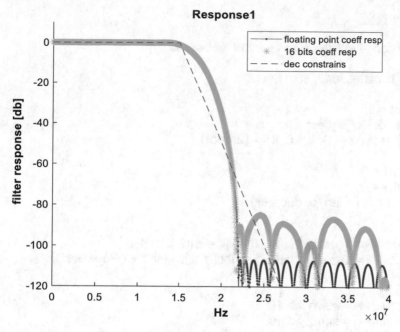

Fig. 3.3 Total frequency response for fixed and floating point FIR implementation

If the designer is satisfied with 80 db stopband rejection he may use fixed point calculations, but unless we gain with respect to processing time, the performance of floating point formats is better.

When dealing with fixed point calculations, the range of sum1, sum2 which is -2^31 to 2^31-1 must be kept, otherwise incorrect results are produced, for our example and 16 bits coefficients this is achieved by limiting the FIR's input to 15 bits signed format.

A fixed point implementation in C which differs between even and odd number of coefficients is

Code 3.1.7

```
void dec_even_fixed(int *x, int n, int dec, int *y)
{
int i, k, count, len2, sum1,sum2;

count = 0;
len2 = 2*len_b_dec;
for (k = (len2 – 1); k < n; k += (2*dec))
 {
 sum1 = 0;
 sum2 = 0;
 for (i = 0; i < len_b_dec; i++)
  {
   sum1 += b_dec[i] * (x[k – i] + x[k – len2 + i + 1]);
   sum2 += b_dec[i] * (x[k – l + dec] + x[k – len2 + i + 1 + dec]);
  }

 y[count++] = sum1 >> 16;
 y[count++] = sum2 >> 16;
 }
}

void dec_odd_fixed(int *x, int n, int dec, int *y)
{
int i, k, count, len2, sum1,sum2;

count = 0;
len2 = 2*len_b_dec – 1;
for (k = (len2 – 1); k < n; k += (2*dec))
 {
 sum1 = 0;
 sum2 = 0;
 for (i = 0; i < (len_b_dec – 1); i++)
  {
   sum1 += b_dec[i] * (x[k – i] + x[k – len2 + i + 1]);
   sum2 += b_dec[i] * (x[k – i + dec] + x[k – len2 + i + 1 + dec]);
  }

 sum1 += b_dec[len_b_dec – 1]*x[k – len_b_dec + 1];
 sum2 += b_dec[len_b_dec – 1]*x[k – len_b_dec + 1 + dec];

 y[count++] = sum1 >> 16;
 y[count++] = sum2 >> 16;
 }
}
```

The floating point coefficients from MATLAB are now stored in the *.h file as int type calculated by the formula

$$b13 = \text{round}\left(b13^* \left(2 \wedge qu - 1\right)\right);$$

The processing time is 77 μS for the 1st filter and 150 μS for the 2nd filter, both for a sampling interval of 60 μS (10000 samples) on an I5 2.9G CPU, not much different from the floating point case.

Using higher than two decimation ratios in one stage is possible, but that does not save processing time. If we keep Ap, Ast and shape factor of the filter constant, then every reduction of 2 in Fp multiplies the number of coefficients, so the processing time for $2 \wedge n$ decimation is about the same for every $n \geq 2$, and is about 305 μS on an I5 2.9G CPU for our 60 μS (10,000 samples) example, relative to 234 μS for 2 X decimation 2 stages.

It is worth to consider one stage FIR decimators if floating point operations are well supported on our CPU, otherwise more and more bits will have to be used to achieve 80 db and more rejection at the stopband.

3.1.5   Higher Decimation Ratio Finite Impulse Response Filters

When we need to decimate by more than 8, a multirate design as declared will be cumbersome, so an efficient design is as follows:

a. Design a CIC decimator by dec/4 where dec is the total decimation.
b. Design a FIR decimator by 4, having the shape factor required by the total response, since this FIR is after the higher decimation, its number of computed outputs is much lower than at the CIC input and using 2 X decimator by 2 instead will not have a significant effect on the total processing time.
c. Correct the droop of the CIC filter in the passband by a suitable compensator.

For comparing to the former example suppose that we need to design a decimator by 16, with final bandwidth of 8 Mhz, following a receiver with IF frequency of 120 Mhz sampled by a 160 Mhz a2d.

3.1.5.1   Cascaded Integrator Comb Decimator Design

Cascaded Integrator Comb (CIC) filter, declared by the Z transform response $1/M^*(1 - Z^M)/(1 - Z^{-1})$ where M is the filter order is described and applied in MATLAB as

Code 3.1.8

```
cic1 = [1 zeros(1,M-1) -1]/M;
a11 = [1 -1];
y = filter(cic1,a11,x);
```

where y,x are the output and input to the filter.

As one stage cannot fulfill the decimation rules above, several stages and/or several *M* type filters are used till the following figure is obtained (Fig. 3.4)

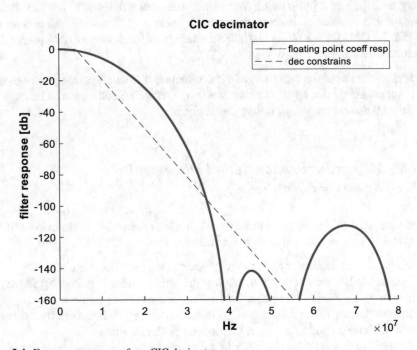

Fig. 3.4 Frequency response for a CIC decimator

The ~0 db response at Fp (4 Mhz) and −100 db at Fstop (36 Mhz) are obtained, a MATLAB code design to do it is

Code 3.1.9

```
res = 2048;
Fs1 = 160e6;
dec1 = 4;
dec2 = 4;

tot_res = Fs1/2/res/dec1/dec2;

F1 = tot_res*(0:res*dec1*dec2 – 1);

b1c = [1 zeros(1,3) -1]/4;
b2c = [1 zeros(1,2) -1]/3;
a11 = [1 -1];
ha1c = freqz(b1c,a11,F1,Fs1);
ha2c = freqz(b2c,a11,F1,Fs1);
ha1 = ha1c.^4.*ha2c.^5;
```

Usually the highest order CIC that meets the decimation rules is about the decimation size, in this case 4 stages of CIC order 4 and 5 stages of CIC order 3 are used in cascade.

The CIC does not use multiples, a fixed point C implementation to the above MATLAB design and 16 signed bits input is

Code 3.1.10

```
void cic_9_stages(short *x, int n, short *y)
{
int k, m, p, offset;
int d[40];
offset = 20;
for (k = 0; k < 40; k++)
  d[k] = 0;

for (k = 0; k < n; k++)
  {
/* 4 stages of CIC1 */
  y[k] = x[k];

  for (m = 0; m < 4; m++)
    {
    d[5*m] = y[k] +  d[5*m + 1];
    y[k] = (d[5*m] – d[5*m + 4]) >> 2;

    for (p = (5*m + 4); p > 5*m; p--)
      d[p] = d[p – 1];
    }

/* 5 stages of CIC2 */
  for (m = 0; m < 5; m++)
    {
    d[offset + 4*m] = y[k] +  d[offset + 4*m + 1];
    y[k] = (d[offset + 4*m] – d[offset + 4*m + 3])/3;

    for (p = (offset + 4*m + 3); p > (offset + 4*m); p--)
      d[p] = d[p – 1];

    }
}
```

The processing time for a verified result is 1.22 mS for a sampling interval of 60 µS on an I5 2.9G CPU, which is much longer than was measured on the low decimation example, since this implementation uses state variables depending on previous states we could not calculate only the decimated outputs, a much better implementation is as follows.

The CIC Z transform $1/M*(1 - Z^{-M})/(1 - Z^{-1})$ is equal to the sum of the geometric sequence $(1, Z^{-1}, Z^{-2} \ldots, Z^{-M})/M$, therefore operating the CIC of order M on the input is equivalent to operating a FIR filter with coefficients $ones(M,1)/M$ on that input.

Since the response of filters in cascade is equal to the response of the convolution of the impulse responses of the individual filters, then the calculation of the equivalent FIR coefficients for the nine stages CIC of order 3 and 4 in MATLAB is as follows:

Code 3.1.11

```
B4 = ones(4,1)/4;
B41 = conv(B4,B4);
B41 = conv(B41,B41);        % 4 stages order 4
B3 = ones(3,1)/3;
B31 = conv(B3,B3);
B31 = conv(B31,B31);
B31 = conv(B31,B3);         % 5 stages order 3
BT = conv(B31,B41);
BT = BT/(sum(BT) + eps);

ha1 = freqz(BT,1,F1,Fs1);
```

That renders 23 coefficients, this FIR implementation can use the dec_odd C function described above with decimation 4 to be processed in 95 μS on an I5 2.9G CPU.

3.1.5.2   Finite Impulse Response Decimator by 4 Design

The above FIR is designed as earlier using trial and error method with the total response and the verification tools, a MATLAB design is given by

Code 3.1.12

```
F2 = tot_res*(0 : (res*dec2 – 1));

dm = fdesign.lowpass('Fp,Fst,Ap,Ast',3.7e6,6e6,0.1,105,Fs1/dec1);
hm = design(dm);
b23 =  hm.numerator;
b23 = b23/(sum(b23) + eps);

ha2 = freqz(b23,1,F2,Fs1/dec1);
hat = ha1(1:length(F2)).*ha2;
```

That renders 77 coefficients for b23, the result is shown as follows with additional zoom on the passband part, where it is seen that the response is not flat as desired (Fig. 3.5)

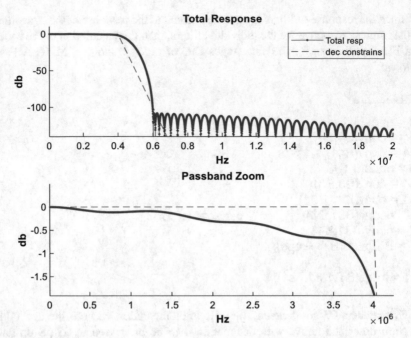

Fig. 3.5 Total frequency response for a non-compensated FIR decimator

3.1.5.3 Compensator Design

A suitable compensator will now be introduced, it is a small FIR of the form [1 a1 a2], being in cascade with the 4 to 1 last decimator, its goal is to keep the attenuation at Fstop to more than 100 db, to keep the attenuation at Fp or less to −1 ± 0.1 db and to keep the ripple at 94% of the passband to minimum, a MATLAB code that does that looks as

Code 3.1.13

```
en1 = min(find(F2 >= 4e6));
en2 = round(0.94*en1);
en3 = min(find(F2 >= 6e6));
mm = 15;

for i1 = -5:0.1:5
  for i2 = -5:0.1:5

  eq = [1 i1 i2 ];
  beq = eq/(sum(eq) + eps);
  h1_eq = freqz(beq,1,F2,Fs1/dec1);
  tr = 20*log10(abs(hat.*h1_eq));
  tr1 = abs(tr(en1));
  tr2 = abs(tr(2:en2));
  tr3 = max(tr(en3:end));
  if (abs(tr1 – 1) < 0.1) & (max(tr2) < mm) & (tr3 < -100)
    i1_opt = i1;
    i2_opt = i2;
    mm = max(tr2);
  end

  end
end
```

After a MATLAB search the optimal compensator is [1 -1.6 -2.7], the integration of the compensator into the MATLAB code is

Code 3.1.14

```
eq = [1 -1.6 -2.7];
beq = eq/(sum(eq) + eps);
h1_eq = freqz(beq,1,F2,Fs1/dec1);
hat_eq = hat.*h1_eq;
b23_eq = conv(b23,eq);
b23_eq = b23_eq/(sum(b23_eq) + eps);
```

Now the response as follows is much better, having a passband ripple of less than 0.1 db, the convolved b23_eq FIR has now 79 coefficients and can use the dec_odd C function described in Sect. 3.1.4 with decimation 4 to be processed in 150 μS on an I5 2.9G CPU for a 60 μS (10,000 samples) input interval (Fig. 3.6).

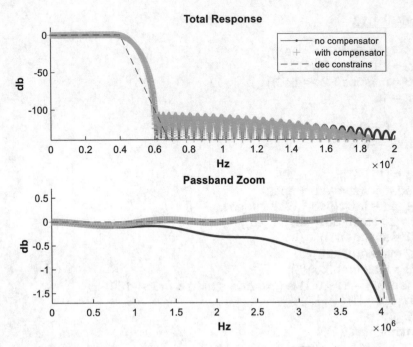

Fig. 3.6 Total response for a compensated versus non-compensated FIR decimator

The last example consumes totally 245 µS processing time for a 60 µS (10,000 samples) input interval relative to 234 µS for the decimator by 4 above.

3.1.6 Conclusions

1. If one needs a high decimation CIC, say 256, then he will need about 4 stages of CIC order 260 and 3 stages of CIC order 180 following the decimation constrains, this structure leads to a 1574 coefficients FIR which is too long although on our I5 2.9G it will run in only 94 µS due to the fact that just N/256 outputs are needed where N is the input length.

2. It would be better to divide it to 2 X 16 CIC decimators, each will contain 4 stages of CIC order 17 and 3 stages of order 12, this structure leads to a 98 coefficients FIR, that can operate even at fixed point 16 bits and together will run in about 116 µS.

3. Regardless of the decimation and number of stages, the processing time on I5 2.9G CPU converges to about 250 µS for 60 µS (10000 samples) time interval at 160 Mhz sampling, meaning that even for a twice faster CPU and 50% spare time, 20–40 Mhz sampling rate for a single channel processed in software is an upper limit for current CPU technology on a single core.

4. As the processing time is relative to the number of filter coefficients and number of computed outputs, then on an I5 2.9G CPU the code in Sect. 3.1.4 uses ~1.6 nS per coefficient per output sample, which is a benchmark for this target to compute FIRs.

3.2 Wide Band Processing

3.2.1 Introduction

In this chapter we describe, demonstrate, and implement shaping windows and FFT, the elements of wide band processing.

3.2.2 Shaping Windows

A shaping window multiplies the digital samples of the a2d by the window coefficients sample by sample, in order to achieve rejection of the neighboring channels (bins) while passing the spectral content of the channel, and is described in MATLAB by

    sigt1 = win. * sig1;

where win is an N size array containing the window coefficients and sig1 is the input signal same size array, if the input is complex then the above process is performed separately on the real and imaginary parts.

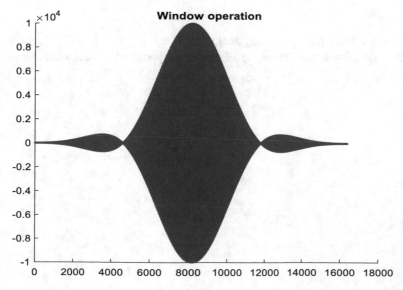

Fig. 3.7 Time response of a window operation on a sine wave

A window operation on a sine wave in time looks like (Fig. 3.7)

There are certain window types which differ by their bandwidth and the amount of rejection of the neighboring channels, a MATLAB code to describe the frequency response of a window for a sampling rate of 204.8 Mhz and an FFT of 4096 points that renders bins of 50 Khz width is

<u>Code 3.2.1</u>

```
Fs = 204.8e6;
N = 4096;
t1 = (0:N-1)/Fs;
win_cheb1 = chebwin(N,106)';
win_cheb1 = win_cheb1/max(win_cheb1);
Fin = 8e6;
res = Fs/N;
Ind = Fin/res + 1;

for k = 1:800

  Fin = 8e6 – 4e5 + k*1000;

  sig1 = 10000*cos(2*pi*Fin*t1 + rand(1));
  sigt = win_cheb1.* sig1;
  Fout = fft(sigt,N)/N;
  Amp_win(k) = abs(Fout(Ind))^2;

end
```

```
Amp_win = Amp_win/max(Amp_win);
```

The following figure shows the frequency response of several windows as follows (Fig. 3.8):

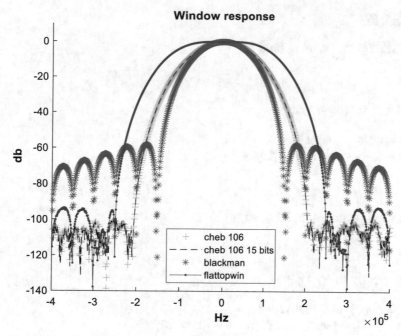

Fig. 3.8 Frequency response of several shaping windows

Since the window is symmetric, just half of the window coefficients are stored in an *.h file for implementation, a useful window is chebwin(N,106) that achieves more than 100 db rejection for float and 15 bits fixed point versions at more than ±3 bins aside.

Floating and fixed point versions for window implementation in C are

Code 3.2.2

```
void win1(float *x, int n, float *y)
{
int i;

for (i = 0; i < (n >> 1); i++)
  {
  y[i] = wol1_win[i]*x[i];
  y[n – 1 – i] = wol1_win[i]*x[n – 1 – i];
  }
}

void win1_fixed(int *x, int n, int *y)
{
int i;

for (i = 0; i < (n >> 1); i++)
  {
  y[i] = (wol1_win[i]*x[i]) >> 15;
  y[n – 1 – i] = (wol1_win[i]*x[n – 1 – i]) >> 15;
  }
}
```

The code exploits the symmetry of the window to calculate 2 outputs for each loop step, the processing time for floating point on an I5 2.9G CPU is 9 μS for 4096 samples and 18 μS for 8192 samples, and for fixed point is 8.5 μS for 4096 samples and 17 μS for 8192 samples, the fixed point window coefficients are computed by wol1_win = round(wol1_win *(2^15 – 1)), the respective *.h file contains half of the window coefficients.

Useful types of windows are window overlap add or WOLA having different orders, a MATLAB code for orders 2–4 is

Code 3.2.3

```
Fs = 204.8e6;
N = 4096;
t2 = (0:2*N-1)/Fs;
win_cheb2 = chebwin(2*N,106)';
win_cheb2 = win_cheb2/max(win_cheb2);
sig2 = 10000*cos(2*pi*Fin*t2 + rand(1));
sigt = win_cheb2.* sig2;
sigt1 = sigt(1:N) + sigt(N + (1:N));
Fout = fft(sigt1,N)/N;
```

For WOLA2 and

```
t3 = (0:3*N-1)/Fs;
win_cheb3 = chebwin(3*N,106)';
win_cheb3 = win_cheb3/max(win_cheb3);
sig3 = 10000*cos(2*pi*Fin*t3 + rand(1));
sigt = win_cheb3.* sig3;
sigt1 = sigt(1:N) + sigt(N + (1:N)) + sigt(2*N + (1:N));
Fout = fft(sigt1,N)/N;
```

For WOLA3 and

```
t4 = (0:4*N-1)/Fs;
win_cheb4 = chebwin(4*N,106)';
win_cheb4 = win_cheb4/max(win_cheb4);
sig4 = 10000*cos(2*pi*Fin*t4 + rand(1));
sigt = win_cheb4.* sig4;
sigt1 = sigt(1:N) + sigt(N + (1:N)) + sigt(2*N + (1:N))+ sigt(3*N + (1:N));
Fout = fft(sigt1,N)/N;
```

For WOLA4

The frequency response of some WOLA order windows is shown for sampling rate of 204.8 Mhz and an FFT of 4096 points (Fig. 3.9)

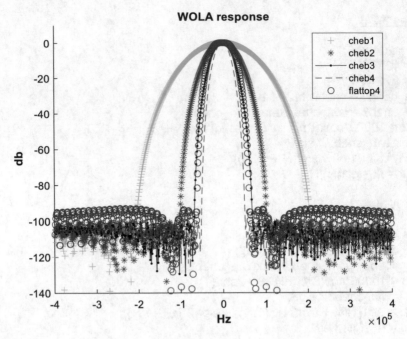

Fig. 3.9 Frequency response of several WOLA windows

When the WOLA order increases, the bandwidth of the channel decreases but the rejection of the neighbors improves so better channel separation is achieved.

A WOLA window which is widely used is WOLA 3 type chebwin(N,106), that rejects the adjacent channel by 36 db and more than 100 db beyond that, floating and fixed point versions for WOLA 3 implementation in C are

Code 3.2.4

```
void wol3(float *x, int dec, float *y)
{
int j, Nfft;
float w1,w2,w3;

Nfft = 8192 / dec;

for (j = 0; j < Nfft; j++)
  {
  w1 = wol3_coeff[j*dec];
  w2 = wol3_coeff[j*dec + 8192];
  w3 = wol3_coeff[j*dec + 2*8192];
  y[j] = w1*x[j] + w2*x[Nfft + j] + w3*x[2*Nfft + j];
  }
}

void wol3_fixed(int *x, int dec, int *y)
{
int j, Nfft;
int w1,w2,w3;

Nfft = 8192 / dec;

for (j = 0; j < Nfft; j++)
  {
  w1 = wol3_coeff[j*dec];
  w2 = wol3_coeff[j*dec + 8192];
  w3 = wol3_coeff[j*dec + 2*8192];
  y[j] = (w1*x[j] + w2*x[Nfft + j] + w3*x[2*Nfft + j]) >> 15;
  }
}
```

The C code uses a 3×8192 elements table and decimation in the table for less 2^n inputs.

The processing time for floating point on an I5 2.9G CPU is 27 µS for 4096 samples and 53 µS for 8192 samples, and for fixed point is 25 µS for 4096 samples and 50 µS for 8192 samples, the fixed point window coefficients are computed by wol3_coeff = round(win_cheb3*($2^{15} - 1$)).

Another useful WOLA window is WOLA 4 type chebwin(N,106), that rejects the adjacent channel by 86 db and more than 100 db beyond that, a floating point version for WOLA 4 implementation in C is

Code 3.2.5

```
void wol4(float *x, int dec, float *y)
{
int j, Nfft;
float w1,w2,w3,w4;

Nfft = 8192 / dec;

for (j = 0; j < Nfft; j++)
  {
  w1 = wol4_coeff[j*dec];
  w2 = wol4_coeff[j*dec + 8192];
  w3 = wol4_coeff[j*dec + 2*8192];
  w4 = wol4_coeff[j*dec + 3*8192];
  y[j] = w1*x[j] + w2*x[Nfft + j] + w3*x[2*Nfft + j] + w4*x[3*Nfft + j];
  }
}
```

The C code uses a 4×8192 elements table and use decimation in the table for less 2^n inputs.

The processing time for floating point on an I5 2.9G CPU is 36 µS for 4096 samples and 72 µS for 8192 samples, in view of the WOLA 3 implementation, the fixed point results would be about the same as for floating point.

3.2.3   Designing Self Made Windows

Window types such as chebwin, flattopwin, hamming, or hanning have certain performance regarding passband shape and neighboring channels rejection, since a window frequency response is equivalent to a FIR filter with the same coefficients, one may design his own window response using the MATLAB fdesign.lowpass function.

The problem with designing long FIR filters is that it takes a lot of time to converge, so we may use a design that has 16 or 32 times less coefficients and then interpolate the result, respectively, to get the final window, the trick is to change slightly the specifications so that (FFT size * WOLA order)/16 will be equal to the length of the WOLA window divided by 16 (32 for both designs applies as well).

Let us modify a WOLA 3 chebwin(106) with fdesign for FFT of 4096

<u>Code 3.2.6</u>

```
Fs = 204.8e6;
N = 4096;
res = 4096*16;
F2 = Fs/2/res*(0:res − 1);

win_cheb3 = chebwin(3*N,106)';
win_cheb3 = win_cheb3 / (sum(win_cheb3) + eps);

dm = fdesign.lowpass('Fp,Fst,Ap,Ast',16*10e3,16*64e3,0.98,105,Fs);
hm = design(dm);
b23 =  hm.numerator;

b23 = interp(b23,16);
b23 = b23 / (sum(b23) + eps);

hai = freqz(b23,1,F2,Fs);
haw = freqz(win_cheb3,1,F2,Fs);
```

The modified 3*N coefficients FIR versus the chebwin(106) window is shown as follows (Fig. 3.10):

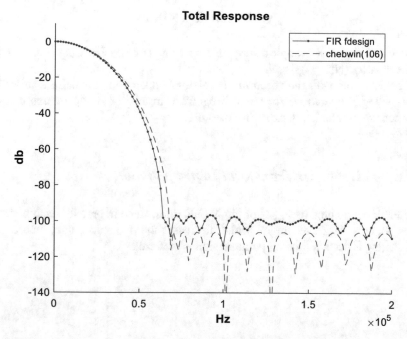

Fig. 3.10 Frequency response of a self-made window versus a known one

3.2.4 Fast Fourier Transform

Fast Fourier transform (FFT) is the tool that renders the complex spectrum of the signals within a certain frequency range with a resolution determined by the FFT size.

MATLAB expression for FFT is

$$y = \text{fft}(x,N)/N$$

which is an algorithm to compute faster the DFT (discrete Fourier transform) expressed by

```
j = sqrt(-1);
for i = 1:N
   y(i) = sum(x.*exp(-j*2*pi*(i – 1)/N*(0:N-1)))/N;
 end
```

where x is a complex or real N points input array and y is a size N complex output array, 1:N/2 elements contain the positive frequencies $(0:N/2 - 1)*Fs/N$ and $(N/2 + 1):N$ elements contain the negative frequencies $(-N/2 : -1)*Fs/N$.

The sequence of exponents $\exp(j*2*pi/N*(0:N-1))$ is called twiddle factors.

The MATLAB expression

$$y = \text{fftshift}(\text{fft}(x,N))/N$$

Renders a size N complex array that contains consecutive frequencies at the interval $(-N/2:(N/2 -1))*Fs/N$.

The FFT computes the spectrum in $\sim N*\log_2 N$ operations for radix 2 algorithm and $\sim N*\log_4 N$ operations for radix 4 algorithm instead of N^2 operations that are required when calculating explicitly the DFT.

3.2.5 Fast Fourier Transform Implementation

The FFT is computed by $\log_2 N$ or $\log_4 N$ butterflies, since in practice each butterfly loses about ½ bit in resolution which decreases the dynamic range, then unless autoscaling is used, a floating point version is preferred.

We will present FFT implementation using decimation in frequency method, where the input sequence is consecutive in time, arranged in complex pairs real(x(0)), imag(x(0)), real(x(1)), imag(x(1)), ... real(x(N − 1)), imag(x(N − 1)), and the complex output sequence should be bit reversed ordered.

The most useful N points FFT implementation performs radix 2 FFT where N is a power of 2, we will present a version that supports 8 to 8192 points, containing the following phases:

3.2.5.1   Bit Reversed Order

Prepare 8192 elements table that contains the bit reversed order data for each output complex element, supporting up to 8192 points FFT, a MATLAB code for it is

code 3.2.7

```
y = 0:8191;
y = dec2bin(y);
y = fliplr(y);
bro = bin2dec(y);
```

The bro array is stored in an *.h file, the step in the table for FFT size N2 is 8192/ N2.

3.2.5.2   Radix 2 Twiddle Factors

Prepare an N/2 complex array that contains the twiddle factors for each N size FFT, a MATLAB code for it is

Code 3.2.8

```
tw_write = [];
%
% 8 points FFT
%
tw = zeros(8,1);
for k = 1:4
  tw(2*k – 1) = cos(2*pi/8*(k – 1));
  tw(2*k) = sin(2*pi/8*(k – 1));
end

y = 0:3;
y = dec2bin(y);
y = fliplr(y);
bro = bin2dec(y) + 1;

for k = 1:length(bro)
  tw_write = [tw_write; tw(2*bro(k) – 1)];
  tw_write = [tw_write; tw(2*bro(k))];
end

%
% 16 points FFT
%
tw = zeros(16,1);
for k = 1:8
  tw(2*k – 1) = cos(2*pi/16*(k – 1));
  tw(2*k) = sin(2*pi/16*(k – 1));
end

y = 0:7;
y = dec2bin(y);
y = fliplr(y);
bro = bin2dec(y) + 1;
```

```
for k = 1:length(bro)
  tw_write = [tw_write; tw(2*bro(k) − 1)];
  tw_write = [tw_write; tw(2*bro(k))];
end
```

....

```
%
% 8192 points FFT
%
tw = zeros(8192,1);
for k = 1:4096
  tw(2*k − 1) = cos(2*pi/8192*(k − 1));
  tw(2*k) = sin(2*pi/8192*(k − 1));
end

y = 0:4095;
y = dec2bin(y);
y = fliplr(y);
bro = bin2dec(y) + 1;

for k = 1:length(bro)
  tw_write = [tw_write; tw(2*bro(k) − 1)];
  tw_write = [tw_write; tw(2*bro(k))];
end
```

Store tw_write to an *.h file, as arranged in bit reversed order.

3.2.5.3   Fast Fourier Transform Radix 2 Implementation

Use a radix 2 FFT implementation from 8 to 8192 points in C [1] as

code 3.2.9

```c
void fft_float(float *x, int n, float *y)
{
short n2, ie, ia, i, j, k, m, step;
float rtemp, itemp, c, s;
const float *w;
int k1, Offset;

/* Calulate start of relevant twiddle factor array (w) */
  Offset = 0;
for (k1 = 8; k1 < n; k1 <<= 1)
  Offset += k1;
w = (const float *)&tw[Offset];

n2 = n;
ie = 1;

for (k = n; k > 1; k >>= 1)
  {
  n2 >>= 1;
  ia = 0;
  for (j = 0; j < ie; j++)
    {
    c = w[2*j];
    s = w[2*j + 1];
    for (i = 0; i < n2; i++)
      {
      m = ia + n2;
      rtemp = c*x[2*m] + s*x[2*m + 1];
      itemp = c*x[2*m + 1] – s*x[2*m];
      x[2*m] = x[2*ia] – rtemp;
      x[2*m + 1] = x[2*ia + 1] – itemp;
      x[2*ia] += rtemp;
      x[2*ia + 1] += itemp;
      ia++;
      }
    ia += n2;
    }
  ie <<= 1;
  }

/* Make bro */
step = 8192/n;
i = 0;
for (k = 0; k < n; k++)
  {
  y[2*k] = x[2*bro[i]];
  y[2*k + 1] = x[2*bro[i] + 1];
  i += step;
  }
}
```

The implementation is in-place, meaning the x array is changed during computation. Processing times for different sizes radix 2 on an I5 2.9G CPU are (Table 3.1) as follows

Table 3.1 Processing times for different sizes radix 2 on an I5 2.9G CPU are

FFT size	Processing time [μS]
128	7
256	16
512	33
1024	70
2048	160
4096	340
8192	700

If fftshift is desired, then the (0:N/2-1) complex elements from y are copied to indexes (N/2:N-1) of output array and (N/2:N-1) complex elements from y are copied to indexes (0 : N/2-1) of output array.

If the power of the FFT bins is desired, then the real$(y(n))^2$ + imag$(y(n))^2$ operation is required and if the phase information is desired, then atan2(imag(y(n)), real(y(n))) is required using a suitable table for the 0:pi/4 range and quadrants conversion later.

For large FFTs use radix 4 implementation to reduce processing time, a 4096 points FFT implementation using decimation in frequency uses also 3 phases as follows:

3.2.5.4 Digit Reversed Order

Prepare 4096 elements table that contains the digit reversed order data for each output complex element, a MATLAB code for it is

code 3.2.10

```
y = 0:4095;
dro = digitrevorder(y,4);
```

3.2.5.5   Radix 4 Twiddle Factors

Prepare an 4096 complex array that contains the twiddle factors for size 4096 FFT, stored in an *.h file and arranged in normal order, the imaginary parts here have negative sign, a MATLAB code for it is

code 3.2.11

```
tw = zeros(4096,1);
for k = 1:4096
  tw(2*k − 1) = cos(2*pi/4096*(k − 1));
  tw(2*k) = -sin(2*pi/4096*(k − 1));
end
```

3.2.5.6   Fast Fourier Transform Radix 4 Implementation

Use a radix 4, 4096 points FFT in C [2] as

code 3.2.12

```c
void fft_radix4_float(float *x, const float *w, short n, float *y)
{
short n1, n2, ie, ia1, ia2, ia3, i0, i1, i2, i3, j, k;
float t, r1, r2, s1, s2, co1, co2, co3, si1, si2, si3;

n2 = n;
ie = 1;
for (k = n; k > 1; k >>= 2)
  {
  n1 = n2;
  n2 >>= 2;
  ia1 = 0;
  for (j = 0; j < n2; j++)
    {
    ia2 = 2*ia1;
    ia3 = ia1 + ia2;
    co1 = w[ia1*2 + 1];
    si1 = w[ia1*2];
    co2 = w[ia2*2 + 1];
    si2 = w[ia2*2];
    co3 = w[ia3*2 + 1];
    si3 = w[ia3*2];
    ia1 += ie;
    for (i0 = j; i0 < n; i0 += n1)
      {
      i1 = i0 + n2;
      i2 = i1 + n2;
      i3 = i2 + n2;
      r1 = x[i0*2] + x[i2*2];
      r2 = x[i0*2] − x[i2*2];

      t = x[2*i1] + x[2*i3];
      x[2*i0] = r1 + t;
      r1 -= t;

      s1 = x[i0*2 + 1] + x[i2*2 + 1];
      s2 = x[i0*2 + 1] − x[i2*2 + 1];

      t = x[2*i1 + 1] + x[2*i3 + 1];
      x[2*i0 + 1] = s1 + t;
      s1 -= t;

      x[2*i2] = r1*co2 + s1*si2;
      x[2*i2 + 1] = s1*co2 − r1*si2;
```

```
        t = x[2*i1 + 1] – x[2*i3 + 1];
        r1 = r2 + t;
        r2 -= t;

        t = x[2*i1] – x[2*i3];
        s1 = s2 – t;
        s2 += t;

        x[i1*2] = co1*r1 + si1*s1;
        x[i1*2 + 1] = co1*s1 – si1*r1;
        x[i3*2] = co3*r2 + si3*s2;
        x[i3*2 + 1] = co3*s2 – si3*r2;
        }
      }
  ie <<= 2;
}

/* Make dro */
for (k = 0; k < n; k++)
  {
  y[2*k] = x[2*dro[k]];
  y[2*k + 1] = x[2*dro[k] + 1];
  }
}
```

The processing time on I5 2.9G CPU is 240 μS relative to 340 μS for the radix 2 version.

3.2.6   *Implementing Large Fast Fourier Transform by Smaller Ones*

When it is necessary to compute a large FFT that cannot be computed directly as explained before, it is possible to do it indirectly by using several smaller FFTs, 3 cases will be shown as follows:

3.2.6.1   Implementing Fast Fourier Transform size N by 16 FFTs Size N/16

Making a 2^{17} points FFT by 16 X 2^{13} points FFT is demonstrated by the following MATLAB code, measuring also the processing times ratio

code 3.2.13

```
j = sqrt(-1);
N = 16*8192;
N16 = 8192;
y = randn(1,N) + j*randn(1,N);
folda = zeros(N16,16);
X = zeros(size(y));

ts16 = tic;

for k = 1:16
  temp = y.*exp(-j*2*pi/N*(k – 1)*[0:N-1]);
  folda(:) = temp;
  X(k:16:16*(N16 – 1) + k) = fft(sum(folda.'));
end

telapsed16 = toc(ts16)

ts = tic;
y1 = fft(y);
telapsed = toc(ts)

telapsed16/telapsed
```

For this case making 16 FFTs of 8192 points is about 23 times slower than direct calculation of a 2^17 points FFT, measured by MATLAB tic-toc pair.

3.2.6.2   Implementing Fast Fourier Transform size N by 4 FFTs size N/4

Making 2^15 points FFT by 4 X 2^13 points FFT is demonstrated by the following MATLAB code, measuring also the processing times ratio

code 3.2.14

```
j = sqrt(-1);
N = 4*8192;
N4 = 8192;
y = randn(1,N) + j*randn(1,N);
folda = zeros(N4,4);
X = zeros(size(y));

for k = 1:4
  temp = y.*exp(-j*2*pi/N*(k – 1)*[0:N-1]);
  folda(:) = temp;
  X(k:4:4*(N4 – 1) + k) = fft(sum(folda.'));
end
```

For this case making 4 FFTs of 8192 points is about 6 times slower than direct calculation of a 2^15 points FFT, measured by MATLAB tic-toc pair.

3.2.6.3   Implementing Fast Fourier Transform Size 3*2^N

Making a 3*2^N points FFT by making 3 FFTs of 2^N points and additional computations is demonstrated by the following MATLAB code, measuring also the processing times ratio

<u>code 3.2.15</u>

```
j = sqrt(-1);
N = 3*8192;
N3 = 8192;
x = randn(1,N) + j*randn(1,N);
Y = zeros(size(x));

x1 = x(1:3:N);
x1_fft = fft(x1);

x2 = x(2:3:N);
x2_fft = fft(x2);

x3 = x(3:3:N);
x3_fft = fft(x3);

W = exp(-j*2*pi/N*[0:N-1]);

for k = 1:8192

  Y(k) = x1_fft(k) + W(k)*x2_fft(k) + W(2*k - 1)*x3_fft(k);

  ind1 = mod(2*k + 16383,24576);
  Y(k + 8192) = x1_fft(k) + W(k + 8192)*x2_fft(k) + W(ind1)*x3_fft(k);

  ind2 = mod(2*k + 32767,24576);
  Y(k + 16384) = x1_fft(k) + W(k + 16384)*x2_fft(k) + W(ind2)*x3_fft(k);

end
```

For this case making 3 FFTs of 8192 points is about 15 times slower than direct calculation of a 3*2^13 points FFT, measured by MATLAB tic-toc pair.

3.2.6.4   Relative Error Analysis

The correctness of the three above algorithms is verified by the relative error in db, calculated by db(abs((Y - ref)./abs(ref)), where ref is the direct computation and Y is the indirect computation.

The relative error for the 3 above algorithms is shown as follows (Fig. 3.11):

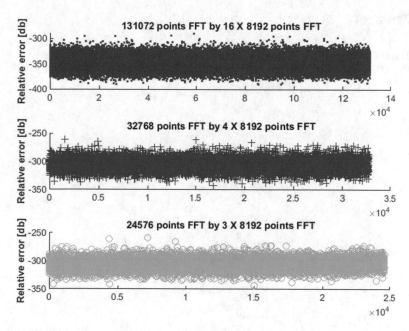

Fig. 3.11   Relative error analysis for 3 types of FFT implementations

3.2.7   Conclusions

In this chapter shaping windows were described and implemented to form the neighboring channels rejection while passing the spectral content of the channel. Later a method to design self made windows was presented, the implementation of radix 2 and radix 4 FFTs was described in details.

Performing large FFTs by smaller ones was demonstrated, in addition there are algorithms that compute 2 X N points real valued FFT using an N points complex FFT, or two different N points real valued FFT using an N points complex FFT, since their benefit is usually for firmware for using less resources, they will not be elaborated here.

References

1. Texas instruments: ftp://ftp.ti.com/pub/tms320bbs/c67xfiles/cfftr2.asm, 1998
2. Burrus CS, Parks TW. DFT/FFT and convolution algorithms and implementation, p. 113

Chapter 4
Complex Finite Impulse Response Filters

Abstract In this chapter we describe and demonstrate the use of complex FIR filters, used for special requirements.

Keywords Complex · FIR · cfirpm · fdesign

4.1 Introduction

In this chapter we describe and demonstrate the use of complex FIR filters, used for special requirements.

4.2 Complex Finite Impulse Response Filter Design

Complex filters are used when non-symmetric passband response is required with respect to 0 Hz (or DC), when a real filter is used, its response is usually computed between 0 and Fs/2 and is symmetric between –Fs/2 and 0, if we use the MATLAB fdesign tool to compute the response of a bandpass filter using the following MATLAB code

Code 4.1

```
Fs = 160e6;
F1 = Fs/8192*(-4096:4095);
dmb =
fdesign.bandpass('Fst1,Fp1,Fp2,Fst2,Ast1,Ap,Ast2',
14.8e6,18.2e6,45.8e6,49.e6,102,0.1,102,Fs);
hmb = design(dmb);
bbp = hmb.Numerator;
h2 = freqz(bbp,1,F1,Fs);
```

© The Author(s), under exclusive license to Springer Nature Switzerland AG 2022
A. Dickman, *Verified Signal Processing Algorithms in MATLAB and C*,
https://doi.org/10.1007/978-3-030-93363-0_4

Which renders a filter with 206 coefficients, and draw its response versus a complex filter with 214 coefficients, we can see that the complex filter passband is just on the positive frequencies (Fig. 4.1).

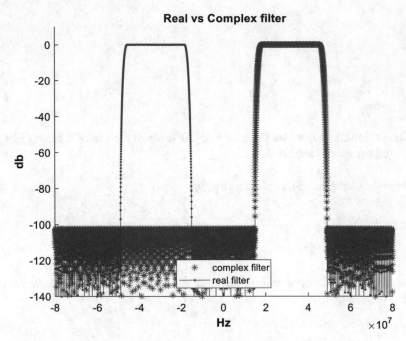

Fig. 4.1 Frequency response of a real versus complex filter

Two methods to design complex FIRs will next be presented

4.2.1 Modified Low Pass Filter Design Method

In this method we design a low pass filter with the fdesign.lowpass tool, and shift the resulting filter to the frequency around which we need, a MATLAB code to do it is

Code 4.2

```
Fs = 160e6;
dm = fdesign.lowpass('Fp,Fst,Ap,Ast',13.9e6,22e6,0.1,102,Fs);
hm = design(dm);
blp = hm.Numerator;
N1 = length(blp);
Fc = 32e6/Fs;

bc = blp.*exp(j*2*pi*Fc*(0:N1-1));
h1 = freqz(bc,1,F1,Fs);
```

This design renders 84 coefficients.

4.2.2   MATLAB Direct Complex Filter Design Method

In this method we use the cfirpm function of MATLAB to get an identical response, a MATLAB code to do it is

Code 4.3

```
f1 = -Fs/2;
f2 = 10.e6;
f3 = 19.6e6;
f4 = 44.2e6;
f5 = 53.8e6;
f6 = Fs/2;

bcom = cfirpm(105,[f1 f2 f3 f4 f5 f6]/(Fs/2),@bandpass);
h2 = freqz(bcom,1,F1,Fs);
```

This design renders 106 coefficients, the @lowpass option may also be used for other cases, but the passband has to include 0 (dc), if we draw both responses on the same figure we get (Fig. 4.2)

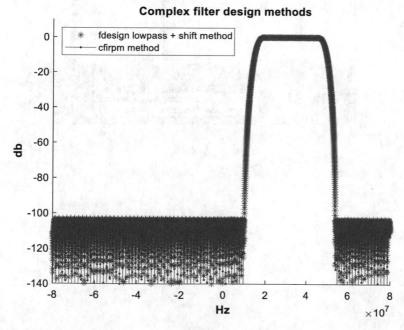

Fig. 4.2 Frequency response of a complex filter for two design methods

4.2.3 Preferred Design Method

In view of more coefficients required for the second method and the necessity to include 0 Hz in the passband at the @lowpass option of the cfirpm, we find the 1st method as preferable.

4.2.4 Property of Complex Filters

The response of the complex conjugate of the filter is the mirror of the complex filter with respect to 0 Hz (dc), namely (Fig. 4.3)

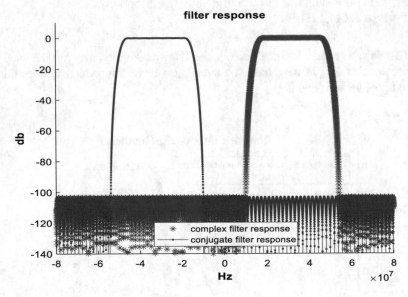

Fig. 4.3 Frequency response of a complex filter versus its conjugate

h1 = freqz(bc,1,F1,Fs);
h1c = freqz(conj(bc),1,F1,Fs);

4.2.5 Implementation of Complex Filters

When multiplying complex numbers, the equation

$(a + ib)*(c + id) = (ac - bd) + i(ad + bc)$

Applies as well to operations with complex filters, and

real(filter(bc,1,xin)) = filter(real(bc),1,real(xin)) − filter(imag(bc),1,imag(xin))
imag(filter(bc,1,xin)) = filter(imag(bc),1,real(xin)) + filter(real(bc),1,imag(xin))

Recall that the implementation of FIRs at Sect. 3.1.4 may be used for complex FIRs by using the real and imaginary parts of the complex filter as separate arrays to implement, the benchmark for FIR implementation at Sect. 3.1.6 was ~1.6 nS per output per coefficient for our target, so if the complex filter operates on real inputs that doubles the benchmark value, and if it operates on complex inputs the benchmark is 4 times higher.

4.3 Conclusions

In this chapter we described and demonstrated the use of complex filters, we recommended on a preferred design method for such a filter and mentioned a useful property of complex filters, later we explained how to implement them in software and the processing time load involved.

A complex filter may serve in a USB or LSB demodulator, which need a non-symmetric response with respect to DC.

Chapter 5
Infinite Impulse Response Filters

Abstract In this chapter the design and implementation of IIR filters are presented, special types of IIR filters are introduced and compared to their compatible FIRs.

Keywords IIR · Difference equations · SOS · Low pass · High pass · Band pass · Notch · All pass

5.1 Introduction

Infinite impulse response (IIR) filters output is a linear combination of the previous outputs and the previous and current inputs, in this chapter their design and implementation will be presented, in addition special types of IIR filters will be introduced and compared to FIRs.

5.2 Infinite Impulse Response Filter Design

Several IIR filter types exist such as elliptic, cheby1, cheby2 and butterworth, a basic IIR design in MATLAB looks as

$$[b,a] = \text{cheby1}(N,R,Wp)$$

where N is the filter's order, R is the passband ripple in db and Wp the passband edge frequency, normalized to ½ of the sampling frequency, [b,a] describe the transfer function, when N = 1 the following equations apply

$$H(z) = Y(Z)/X(Z) = \left(b(1)+b(2)*Z^{-1}\right)/\left(1+a(2)*Z^{-1}\right) \tag{5.1}$$

Implemented by the direct form 1 difference equations

$$D(z) = X(z)/\left(1+a(2)*Z^{-1}\right) \text{ or}$$

$$d(n) = x(n) - a(2)*d(n-1)$$

$$y(n) = b(1)*d(n) + b(2)*d(n-1)$$

Implementing such a filter in MATLAB is

Code 5.1

```
Rp = 1;
Rs = 60;
Wp = 0.2;
[b,a] = ellip(1,Rp,Rs,Wp);
b = b*sum(a)/sum(b);
```

To implement in C for floating type we store the a(1),b(1),b(2) sequence in an *.h file, and for fixed point 16 bits quantization we store the coefficients as round(coeff*(2^15 − 1)) in the same order, then perform

Code 5.2

```
void iir_1st_float(float *x, const float *w, int n, float *y)
{
int k;
float d[2];

d[0] = (float)0.;
d[1] = (float)0.;

for (k = 0; k < n; k++)
  {
  d[0] = x[k] − w[0]*d[1];
  y[k] = w[1]*d[0] + w[2]*d[1];
  d[1] = d[0];
  }
}

void iir_1st_fixed(int *x, const int *w, int n, int *y)
{
int k, d[2];

d[0] = 0;
d[1] = 0;

for (k = 0; k < n; k++)
  {
  d[0] = x[k] − ((w[0]*d[1]) >> 15);
  y[k] = (w[1]*d[0] + w[2]*d[1]) >> 15;
  d[1] = d[0];
  }
}
```

The processing time for floating point on an I5 2.9G CPU is 32 nS per sample and for fixed point 30 nS per sample.

Code 5.2 may be used for lead or lag compensators used in control loops to stabilize the loop, by effecting phase and gain margins.

When N = 2 then the following equations apply

$$H(z) = Y(Z)/X(Z) = \left(b(1) + b(2)*Z^{-1} + b(3)*Z^{-2}\right)/\left(1 + a(2)*Z^{-1} + a(3)*Z^{-2}\right) \quad (5.2)$$

Implemented by the direct form 2 equations

$$D(z) = X(z)/\left(1 + a(2)*Z^{-1} + a(3)*Z^{-2}\right) \text{or}$$

$$d(n) = x(n) - a(2)*d(n-1) - a(3)*d(n-2)$$

$$y(n) = b(1)*d(n) + b(2)*d(n-1) + b(3)*d(n-2)$$

Or second order section (SOS) in MATLAB terminology.

For designing higher order IIRs we use the MATLAB fdesign tool again, this time using IIRs, which considers the specifications and divides the design into a group of second order sections (SOS), this solution minimizes the effect of fixed point quantization and noise on the filter's response, which may even lead to instability when implementing directly high order filters.

5.2.1 Low Pass Filter Design

We will design and compare a low pass IIR filter design with identical specifications to a FIR design.

A MATLAB code that includes both designs is

Code 5.3

```matlab
Fs = 160e6;
F1 = Fs/2/4096*(0:4095);

% FIR design
dm = fdesign.lowpass('Fp,Fst,Ap,Ast',6.97e6,11e6,0.1,103,Fs);
hm = design(dm);   % 168 coeff
b13 =  hm.numerator;
b13 = b13/(sum(b13) + eps);
h_fir = freqz(b13,1,F1,Fs);

% IIR design
dm = fdesign.lowpass('Fp,Fst,Ap,Ast',6.97e6,11e6,0.1,103,Fs);
f = design(dm,'ellip');

b1 = f.sosMatrix(1,1:3);
a1 = f.sosMatrix(1,4:6);
b1 = b1*sum(a1)/sum(b1);

b2 = f.sosMatrix(2,1:3);
a2 = f.sosMatrix(2,4:6);
b2 = b2*sum(a2)/sum(b2);

.
.
.


b5 = f.sosMatrix(5,1:3);
a5 = f.sosMatrix(5,4:6);
b5 = b5*sum(a5)/sum(b5);

h1 = freqz(b1,a1,F1,Fs);
h2 = freqz(b2,a2,F1,Fs);
h3 = freqz(b3,a3,F1,Fs);
h4 = freqz(b4,a4,F1,Fs);
h5 = freqz(b5,a5,F1,Fs);

h = h1.*h2.*h3.*h4.*h5;
```

The FIR design renders 168 coefficients for b13 and the IIR design renders 5 second order sections (SOS or biquads), the "ellip" option is generally the most efficient one, respective responses h_fir and h are shown as follows (Fig. 5.1).

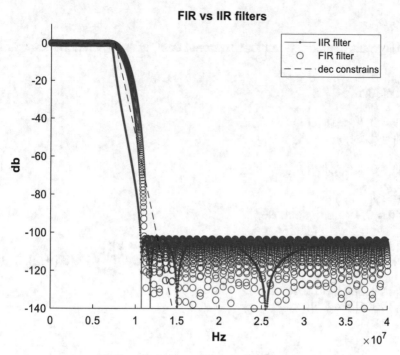

Fig. 5.1 Frequency response of FIR versus IIR low pass filters

To implement the IIR design in C with floating point types we store the SOS floating point coefficients in an *.h file, 5 per section and ordered as a1(2), a1(3), b1(1), b1(2), b1(3), ..., a5(2), a5(3), b5(1), b5(2), b5(3), totally 25 coefficients, the code allocates 3 state variables per section, totally up to 10 sections and a C code is

Code 5.4

```c
void iir_multi_section_float(float *x, const float *w, int n, int num_sect, float *y)
{
int i,k;
float d[30];

for (k = 0; k < 30; k++)
  d[k] = (float)0.;

for (k = 0; k < n; k++)
  {
  y[k] = x[k];
  for (i = 0; i < num_sect; i++)
    {
    d[3*i] = y[k] – w[5*i]*d[3*i + 1] – w[5*i + 1]*d[3*i + 2];
    y[k] = w[5*i + 2]*d[3*i] + w[5*i + 3]*d[3*i + 1] + w[5*i + 4]*d[3*i + 2];

    d[3*i + 2] = d[3*i + 1];
    d[3*i + 1] = d[3*i];
    }
  }
}
```

For fixed point implementation the coefficients are stored in an *.h file as round(coeff*($2^{10} - 1$)) as only 11 bits are required to keep the filter's specifications, and a C code is

Code 5.5

```
void iir_multi_section_fixed(int *x, const int *w, int n, int num_sect, int shy, int *y)
{
int i,k,d[30];

for (k = 0; k < 30; k++)
 d[k] = 0;

for (k = 0; k < n; k++)
 {
 y[k] = x[k];

 for (i = 0; i < num_sect; i++)
  {
  switch (i)
   {
   case 0:
    y[k] >>= 2;
    break;
   case 1:
    y[k] >>= 2;
    break;
   case 2:
    y[k] >>= 1;
    break;
   }

  d[3*i] = y[k] − ((w[5*i]*d[3*i + 1] + w[5*i + 1]*d[3*i + 2]) >> coeff_shift);
  y[k] = (w[5*i + 2]*d[3*i] + w[5*i + 3]*d[3*i + 1] + w[5*i + 4]*d[3*i + 2]);
  y[k] >>= shy;

  d[3*i + 2] = d[3*i + 1];
  d[3*i + 1] = d[3*i];
  }
 }
}
```

Where coeff_shift is defined in the *.h file and is 10 in this case, the switch is added here as every section's output is amplified by up to 1/sum(bi), so y[k] has to be attenuated at some of the sections so that the int limits are not violated, the amount of right shift (attenuation) may be found by emulating the iir_multi_section_fixed function in MATLAB to find the shift that prevents the display of "int violation" as follows:

Code 5.6

```
Fs = 160e6;
d = zeros(15,1);
N = 10000;
t = (0:N-1)/Fs;
qu1 = 10;
coff = [a1(2:end) b1 a2(2:end) b2 a3(2:end) b3 a4(2:end) b4 a5(2:end) b5];
coff = round(coff*(2^qu1 – 1));
que = 9;

for i = 1:1:80
 f1 = i*1e6;
 x = round(32767*sin(2*pi*f1*t));

 for k = 1:N
  y(k) = x(k);
   for m = 1:5

     switch m
      case 1
        y(k) = y(k)/4;
      case 2
        y(k) = y(k)/4;
      case 3
        y(k) = y(k)/2;
     end

    d(3*(m-1) + 1) = -coff(5*(m-1)+1)*d(3*(m-1)+2) –
    coff(5*(m-1)+2)*d(3*(m-1)+3);

    if ((y(k) + d(3*(m-1)+1)) > (2^31 – 1))
       display('int violation')
    end

    d(3*(m-1)+1) = y(k) + round(d(3*(m-1)+1)/2^qu1);

    y(k) = coff(5*(m-1)+3)*d(3*(m-1)+1) + coff(5*(m-1)+4)*d(3*(m-1)+2)+...
      coff(5*(m-1)+5)*d(3*(m-1)+3);

    if (y(k) > (2^31 – 1))
       display('int violation')
    end

    y(k) = round(y(k)/2^que);

    d(3*(m-1)+3) = d(3*(m-1)+2);
    d(3*(m-1)+2) = d(3*(m-1)+1);
   end
 end
end
```

Processing times per sample results on an I5 2.9G CPU are (Table 5.1) as follows

Table 5.1 Processing times per sample results on an I5 2.9G CPU are

Processing time per sample	Floating point [nS]	Fixed point [nS]
1 section	36	39
2 sections	47	52
3 sections	58	66
4 sections	69	79
5 sections	79	91

And 61 nS per sample for the competing FIR in floating point format on the same I5 2.9G CPU, using code 3.1.6 in Sect. 3.1.4, since the FIR is computed only on the decimated outputs then processing time is saved relative to the IIR where all input samples participate.

5.2.1.1   Sensitivity to Quantization of Coefficients

The 168 coefficients FIR needs at least 23 bits quantization of the form round $(b13*(2^{qu} - 1))/(2^{qu} - 1)$ in order to maintain the specifications, while for the IIR implementation 11 bits for the five sections are adequate.

5.2.1.2   Group Delay Assessment

The group delay of a filter expresses the delay of the filter versus frequency, and for both filters is calculated and plotted as follows (Fig. 5.2):

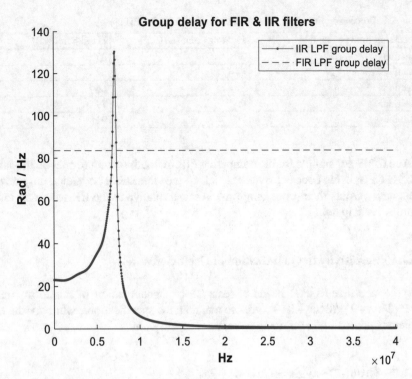

Fig. 5.2 Group delay of FIR versus IIR low pass filters

Code 5.7

```
[gdf,wdf] = grpdelay(b13,2048);         % FIR group delay
wdf = wdf/pi*Fs/2;

[gdi,wdi] = grpdelay(f.sosMatrix,2048); % IIR group delay
wdi = wdi/pi*Fs/2;
```

The group delay for a linear phase FIR is about ½ of the number of coefficients in samples, which means a constant delay for all frequencies and some applications require that as shown the group delay of an IIR filter is not constant.

5.2.1.3 Comparing Low Pass Infinite Impulse Response Filter & FIR

(a) The processing time for same specifications filters is about the same time as for IIR, only the decimated outputs are computed for the FIR.
(b) The sensitivity to quantization of coefficients is better for IIRs, especially when long FIRs are involved.

(c) If constant group delay is a must, then FIR implementation would be preferable.
(d) If a fixed point IIR implementation is chosen, the designer should verify that the
limits of int type are not violated as shown in code 5.6.

5.2.2   High Pass Filter Design

We will design and compare a high pass IIR filter design with identical specifications to a FIR design.

A MATLAB code that includes both designs is

Code 5.8

```
Fs = 20e6;
F1 = Fs/2/4096*(0:4095);

% Fir design
dm = fdesign.highpass( 'Fst,Fp,Ast,Ap',0.8e6,1.2e6,80,1,Fs);
hm = design(dm);   % 139 coeff
b13 =  hm.numerator;
h_fir = freqz(b13,1,F1,Fs);

% IIR design
dm = fdesign.highpass( 'Fst,Fp,Ast,Ap',0.8e6,1.2e6,80,1,Fs);
f = design(dm,'ellip');

b1 = f.sosMatrix(1,1:3);
a1 = f.sosMatrix(1,4:6);

.
.
.


b4 = f.sosMatrix(4,1:3);
a4 = f.sosMatrix(4,4:6);

h1 = freqz(b1,a1,F1,Fs);
h2 = freqz(b2,a2,F1,Fs);
h3 = freqz(b3,a3,F1,Fs);
h4 = freqz(b4,a4,F1,Fs);

G = prod(f.ScaleValues);
h = G*h1.*h2.*h3.*h4;
```

The FIR design renders 139 coefficients for b13 and the IIR design renders 4 second order sections (SOS or biquads), the "ellip" option is generally the most efficient one, respective filters responses h_fir and h are as follows (Fig. 5.3).

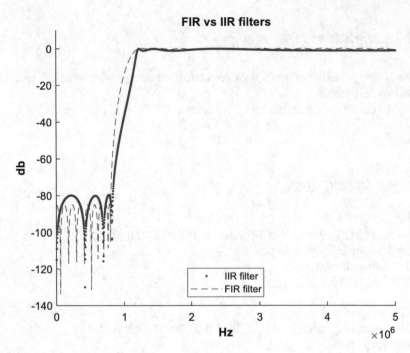

Fig. 5.3 Frequency response of FIR versus IIR high pass filters

To implement the IIR design in C with floating point types we store the SOS floating point coefficients in an *.h file, 5 per section and ordered as a1(2), a1(3), b1(1), b1(2), b1(3), ..., a4(2), a4(3), G*b4(1), G*b4(2), G*b4(3), totally 20 coefficients, the code allocates 3 state variables per section, totally up to 10 sections and the C code is exactly as Code 5.4 above for floating point except that num_sect is 4.

For fixed point implementation the coefficients are stored in an *.h file as round(coeff*(2^11 − 1)) as only 12 bits are required to keep the filter's specifications, and a C code is exactly as Code 5.5 above, where coeff_shift is defined in an *.h file and is 11 in this case, the switch cases are the same as for the IIR LPF, and the cases values are found by an emulation as in Code 5.6.

The processing times per sample results on an I5 2.9G CPU are the same as in Table 5.1 above, and 228 nS per sample for the competing FIR in floating point format on the same I5 2.9G CPU, using code 3.1.6 in Sect. 3.1.4.

5.2.2.1   Sensitivity to Quantization of Coefficients

The 139 coefficients FIR needs at least 18 bits quantization of the form round(b13*(2^qu – 1))/(2^qu – 1) in order to maintain the specifications, while for the IIR implementation 12 bits for the four sections are adequate.

5.2.2.2   Group Delay Assessment

The group delay for both filters is calculated as in Code 5.7 above and plotted as follows (Fig. 5.4):

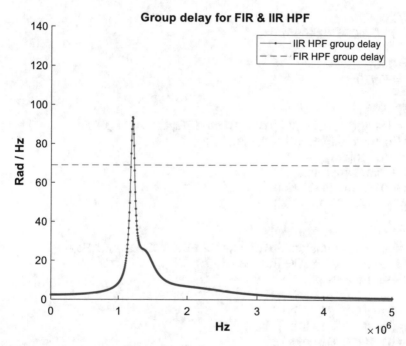

Fig. 5.4 Group delay of FIR versus IIR high pass filters

5.2.2.3   Comparing High Pass Infinite Impulse Response Filter & FIR

(a) The processing time for same specifications filters is 3-4 times shorter for IIR, decimation is not performed for HPF.
(b) The sensitivity to quantization of coefficients is again better for IIRs, especially when long FIRs are involved.
(c) If constant group delay is a must, then FIR implementation will be preferable, otherwise IIR will consume less resources and be less sensitive to coefficients quantization.
(d) If a fixed point IIR implementation is chosen, the designer should verify that the limits of int type are not violated as shown in code 5.6.

5.2.3 Band Stop Filter Design

Band stop or notch filter rejects a narrow frequency band while passing other frequencies, it is usually used for attenuating a disturbance at specific frequency, or attenuating mechanical resonances in control loops, where a solution of redesigning the mechanics would be expensive and time consuming.

We will design and compare a notch IIR filter design with identical specifications to a FIR design, a MATLAB code that includes both designs is

Code 5.9

```
Fs = 20e6;
F1 = Fs/2/4096*(0:4095);

Fc = 6e6;
BW = Fc/20;

% Fir design
dm = fdesign.bandstop('Fp1,Fst1,Fst2,Fp2,Ap1,Ast,Ap2',Fc –
BW,Fc,Fc+100,Fc+BW+100,1,80,1,Fs);  % 185 coeff
hm = design(dm);
b13 =  hm.numerator;
b13 = b13/(sum(b13) + eps);
h_fir = freqz(b13,1,F1,Fs);

% IIR design with same spec
dm = fdesign.bandstop('Fp1,Fst1,Fst2,Fp2,Ap1,Ast,Ap2',Fc –
BW,Fc,Fc+100,Fc+BW+100,1,80,1,Fs);
f = design(dm,'ellip');

b1 = f.sosMatrix(1,1:3);
a1 = f.sosMatrix(1,4:6);
b1 = b1*sum(a1)/sum(b1);

b2 = f.sosMatrix(2,1:3);
a2 = f.sosMatrix(2,4:6);
b2 = b2*sum(a2)/sum(b2);

b3 = f.sosMatrix(3,1:3);
a3 = f.sosMatrix(3,4:6);
b3 = b3*sum(a3)/sum(b3);

h1 = freqz(b1,a1,F1,Fs);
h2 = freqz(b2,a2,F1,Fs);
h3 = freqz(b3,a3,F1,Fs);

h = h1.*h2.*h3;
```

The -3db corner frequency aside from Fc is denoted as BW and represents ½ of the filter's bandwidth, increasing the ratio between Fc and BW reduces the phase delay at low frequencies, which is an advantage for control loops which need to keep enough phase margin, provided that the uncertainty in resonance frequency is such that the filter's rejection can handle.

The FIR design renders 185 coefficients for b13 and the IIR design renders 3 second order sections (SOS or biquads), the "ellip" option is generally the most efficient one, respective responses are shown as follows (Fig. 5.5).

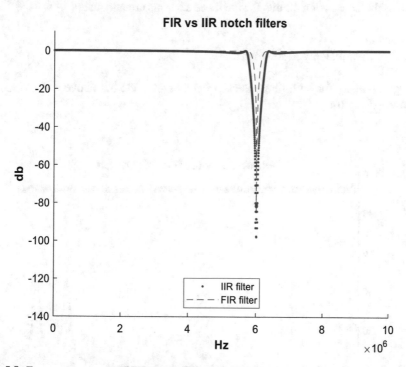

Fig. 5.5 Frequency response of FIR versus IIR notch filters

To implement the IIR design in C with floating point types we store the SOS floating point coefficients in an *.h file, 5 per section and ordered as a1(2), a1(3), b1(1), b1(2), b1(3), ..., a3(2), a3(3), b3(1), b3(2), b3(3), totally 15 coefficients, the code allocates 3 state variables per section, totally up to 10 sections and the C code is exactly as Code 5.4 above for floating point except that num_sect is 3.

For fixed point implementation the coefficients are stored in an *.h file as round(coeff*(2^11 − 1)) as only 12 bits are required to keep the filter's specifications, and a C code is exactly as Code 5.5 above, where coeff_shift is defined in an *.h file and is 11 in this case, the switch cases are not required as 1/sum(bi) for an IIR notch is smaller than 1.

The processing times per sample results on I5 2.9G CPU are the same as in Table 5.1 above, and 303 nS per sample for the competing FIR in floating point format on the same I5 2.9G CPU, using code 3.1.6 in Sect. 3.1.4.

5.2.3.1   Sensitivity to Quantization of Coefficients

The 185 coefficients FIR needs at least 16 bits quantization of the form round(b13*(2^qu – 1))/(2^qu – 1) in order to maintain the specifications, while for the IIR implementation 12 bits for the three sections are adequate.

5.2.3.2   Group Delay Assessment

The group delay for both filters is calculated as in Code 5.7 above and plotted as follows (Fig. 5.6):

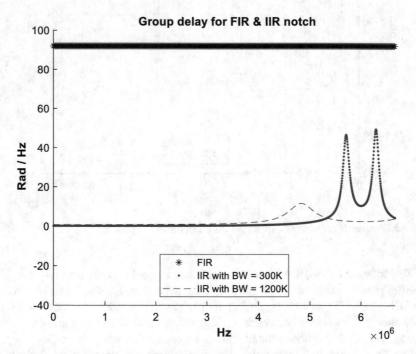

Fig. 5.6 Group delay of FIR versus IIR notch filters

5.2.3.3   Comparing Band Stop Infinite Impulse Response Filter & FIR

(a) The processing time for same specifications filters is about 5 times shorter for IIR, decimation is not performed for notch filters.
(b) The sensitivity to quantization of coefficients is again better for IIRs, especially when long FIRs are involved.
(c) when constant group delay is not required for notch filters, IIR implementation is preferable.

5.2.4   All Pass Filter Design

A filter with 0 db gain response and a monotonically decreasing phase at the range [0 Fs/2] is an all pass filter, the equations of such a filter are

$$H(z) = Y(Z)/X(Z) = (a + Z^{-n1})/(1 + a * Z^{-n1}) \tag{5.3}$$

Implemented by the direct form 1 difference equations

$$D(z) = X(z)/(1 + a * Z^{-n1}) \text{ or}$$

$$d(n) = x(n) - a * d(n - n1)$$

$$y(n) = a * d(n) + d(n - n1)$$

Implementing such a filter in MATLAB is

<u>Code 5.10</u>

```
Fs = 16e6;
F1 = Fs/8192*(0:4095);
a = -0.05;
n1 = 2;

a11 = [1  zeros(1,n1)  a];
b11 = [a  zeros(1,n1)  1];

h1 = freqz(b11,a11,F1,Fs);
```

Usually a is between -1 and 0, the phase characteristics depend on a and n1, when plotting the gain and phase we get (Fig. 5.7)

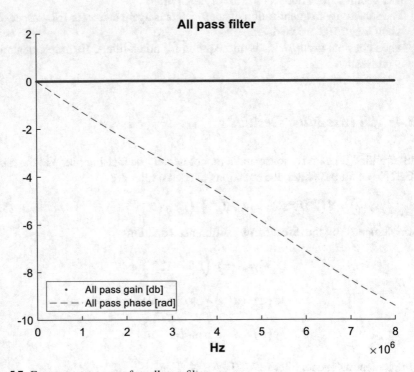

Fig. 5.7 Frequency response of an all pass filter

A C implementation with fixed point format is as follows, where a is round(-0.05*32767) or -1638

Code 5.11

```
void all_pass(short *x, int n, int n1, short a, short *y)
{
int k, p;
int d[10];

for (k = 0; k < 10; k++)
  d[k] = 0;

for (k = 0; k < n; k++)
  {
  d[0] = x[k] - ((a*d[n1]) >> 15);
  y[k] = ((a*d[1]) >> 15) + d[n1];

  for (p = n1; p > 0; p--)
    d[p] = d[p − 1];
  }
}
```

The processing time per sample on I5 2.9G CPU is 78 nS.

5.2.5   Band Pass Filter Design

Band pass filter passes a narrow frequency band while rejecting all other frequencies, we will design and compare a band pass IIR filter design with identical specifications to a FIR design, a MATLAB code that includes both designs is

Code 5.12

```
Fs = 20e6;
Fc = 6e6;
BW = 300e3;
F1 = Fs/2/4096*(0:4095);

% Fir design
dm = fdesign.bandpass('Fst1,Fp1,Fp2,Fst2,Ast1,Ap,Ast2',Fc - 2*BW,Fc -
BW,Fc + BW,Fc + 2*BW,80,0.2,80,Fs);   % 227 coeff
hm = design(dm);
b13 =  hm.numerator;
h_fir = freqz(b13,1,F1,Fs);

% IIR design with same spec
dm = fdesign.bandpass('Fst1,Fp1,Fp2,Fst2,Ast1,Ap,Ast2',Fc - 2*BW,Fc -
BW,Fc + BW,Fc + 2*BW,80,0.2,80,Fs);
f = design(dm,'ellip');

b1 = f.sosMatrix(1,1:3);
a1 = f.sosMatrix(1,4:6);

b2 = f.sosMatrix(2,1:3);
a2 = f.sosMatrix(2,4:6);

b3 = f.sosMatrix(3,1:3);
a3 = f.sosMatrix(3,4:6);

b4 = f.sosMatrix(4,1:3);
a4 = f.sosMatrix(4,4:6);

b5 = f.sosMatrix(5,1:3);
a5 = f.sosMatrix(5,4:6);

b6 = f.sosMatrix(6,1:3);
a6 = f.sosMatrix(6,4:6);

b7 = f.sosMatrix(7,1:3);
a7 = f.sosMatrix(7,4:6);

h1 = freqz(b1,a1,F1,Fs);
h2 = freqz(b2,a2,F1,Fs);
h3 = freqz(b3,a3,F1,Fs);
h4 = freqz(b4,a4,F1,Fs);
h5 = freqz(b5,a5,F1,Fs);
h6 = freqz(b6,a6,F1,Fs);
h7 = freqz(b7,a6,F1,Fs);

G = prod(f.ScaleValues);

h = G*h1.*h2.*h3.*h4.*h5.*h6.*h7;
```

The FIR design renders 227 coefficients for b13 and the IIR design renders 7 second order sections (SOS or biquads), the "ellip" option is generally the most efficient one, respective responses are shown as follows.

As can be seen from Fig. 5.8, the stop band boundaries are narrower than the FIR ones, so the IIR design is modified as follows:

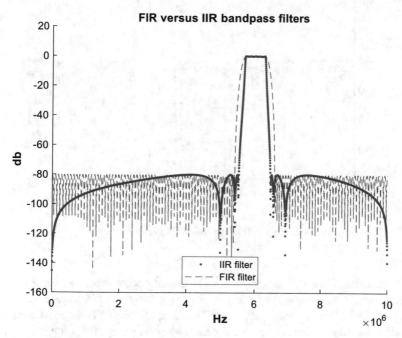

Fig. 5.8   Frequency response of FIR versus IIR bandpass filters

Code 5.13

```
dm = fdesign.bandpass('Fst1,Fp1,Fp2,Fst2,Ast1,Ap,Ast2',Fc - 2.2*BW,Fc -
BW,Fc + BW,Fc + 2.2*BW,80,0.2,80,Fs);
f = design(dm,'ellip');
```

This IIR design renders 6 second order sections (SOS or biquads), with the following response which now complies with the FIR design (Fig. 5.9)

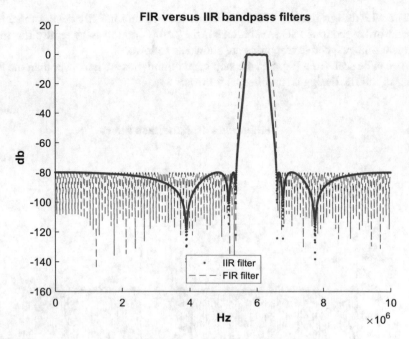

Fig. 5.9 Frequency response of FIR versus modified IIR bandpass filters

To implement the IIR design in C with floating point types we store the SOS floating point coefficients in an *.h file, 5 per section and ordered as a1(2), a1(3), b1(1), b1(2), b1(3), ..., a6(2), a6(3), b6(1), b6(2), b6(3), totally 30 coefficients, the C code is exactly as Code 5.4 above for floating point except that num_sect is 6.

For fixed point implementation the coefficients are stored in an *.h file as round(coeff*(2^10 − 1)) as only 11 bits are required to keep the filter's specifications, and a C code is exactly as Code 5.5 above, where coeff_shift is defined in an *.h file and is 10 in this case, the switch cases are not required as 1/sum(bi) for an IIR bandpass is smaller than 1.

The processing time for 6 SOS on I5 2.9G CPU is estimated from Table 5.1 as 90 nS per sample, and 372 nS per sample for the competing FIR in floating point format on the same I5 2.9G CPU, using code 3.1.6 in Sect. 3.1.4.

5.2.5.1   Sensitivity to Quantization of Coefficients

The 227 coefficients FIR needs at least 20 bits quantization of the form round(b13*(2^qu − 1))/(2^qu − 1) in order to maintain the specifications, while for the IIR implementation 11 bits for the six sections are adequate.

5.2.5.2   Group Delay Assessment

The group delay for both filters is calculated as in Code 5.7 above and plotted as follows (Fig. 5.10):

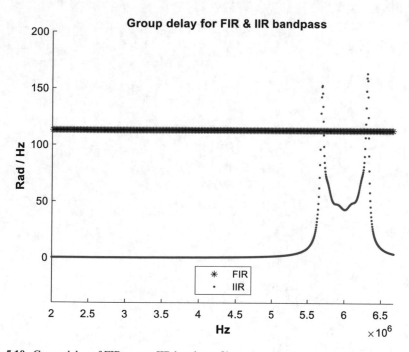

Fig. 5.10   Group delay of FIR versus IIR bandpass filters

5.2.5.3   Comparing Band Pass Infinite Impulse Response Filter & FIR

(a) The processing time for same specifications filters is about 4 times shorter for IIR, decimation is not performed for bandpass filters.
(b) The sensitivity to quantization of coefficients is again better for IIRs, especially when long FIRs are involved.
(c) If Constant group delay is not required for bandpass filters, IIR implementation is preferable.

5.3   Conclusions

In this chapter design and implementation of IIR filters were presented and compared to FIR filters, several types of IIR filters were introduced and investigated.

Chapter 6
Non-Linear and Batch Filters

Abstract The non-linear median filter is described and compared to a batch filter which performs forward and reverse filtering.

Keywords Non-linear · Median · Batch filter · Filtfilt

6.1 Introduction

The filters FIR, IIR, or CIC are linear, as follows a median filter which is a non-linear operation and the reasons to use it will be described, the filter will be compared to a batch filter which performs forward and reverse filtering with minimal transient time.

6.2 Median Filter

When the input signal contains noise, spikes, or interruptions, the intuitive solution is to use a filter such as FIR, IIR, or CIC to mitigate those unwanted disturbances, however linear filtering is usually accompanied by a delay which may be intolerable, which for a FIR it is about ½ the filter length samples, the lower the filter's pass band frequency, the greater the delay.

In addition, if the pass band zone is defined, using "stronger" filtering would violate that requirement and attenuate frequencies that should not be attenuated.

Therefore a non-linear filter may be used, such as a median filter, described by the following MATLAB code:

© The Author(s), under exclusive license to Springer Nature Switzerland AG 2022   89
A. Dickman, *Verified Signal Processing Algorithms in MATLAB and C*,
https://doi.org/10.1007/978-3-030-93363-0_6

Code 6.1

```
y = zeros(1,length(sig));
med = 2;
for k = (med + 1) : (length(sig) – med)
  y(k) = median(sig((k-med) : (k+med)));
end
```

where sig is the noisy signal, y is the filtered signal and (2*med + 1) is the median length, the filter has med samples delay and in the average no delay, the median length should be about 10% of the highest expected frequency cycle in samples in order to maximize the filtering and still not distort the signal.

A C code implementation of a median filter is as follows:

Code 6.2

```
void median_filt(int *x, int n, int med, int *y)
{
int i, j, k, ind, st, c1, c2, i1, max1;
int temp_arr[33];

c1 = 0;
for (i = med; i < (n – med); i++)
  {
  ind = med + 1;
  st = i – med;
c2 = 0;
for (j = st; j < (i + med + 1); j++)
  temp_arr[c2++] = x[j];

for (k = 0; k < ind; k++)
  {
  max1 = 0x80000000;
  i1 = 0;

for (j = 0; j < c2; j++)
  {
  if (temp_arr[j] > max1)
    {
    max1 = temp_arr[j];
    i1 = j;
    }
  }

temp_arr[i1] = 0x80000000;
}

y[c1++] = max1;

}
}
```

The code calculates the median of the array x[(k – med) : (k + med)] up to med = 16, it ignores med samples on each side of the input array x for simplification, the processing time per sample on an I5 2.9G CPU is 31 nS for med = 1, 110 nS for med = 3 and 302 nS for med = 5.

6.3   Batch Filter with Comparison to Median Filter

A batch filter is a filter that is used when a batch of samples is to be filtered and is long enough to enable forward and reverse filtering, the MATLAB function filtfilt() performs such a filter provided that the number of samples is at least three times the filter order, appropriate initial conditions are applied to the filter to minimize transient response.

The filtfilt function has no delay and its noise rejection in db is double than filter() with the same parameters due to double filtering operation.

In the following table a white Gaussian noisy input with different input frequencies will be tested.

Table 6.1 Relative error performances for different filters

Input frequency	Med order	Relative error for median [db]	Relative error for FIR filter [db]	Relative error for FIR filtfilt [db]
Fs/50	5	−20.3	−22.2	−24
Fs/20	2	−17.8	−15.4	−25.2
Fs/10	1	−17.2	−22.6	−22.4

The filters used for Table 6.1 are FIRs whose Fp is the input frequency, and the associated delay is ½ the number of the filter's length, the median has med samples delay and filtfilt() has no delay, for an input frequency of Fs/50 the filters output is shown in the following figure (Fig. 6.1):

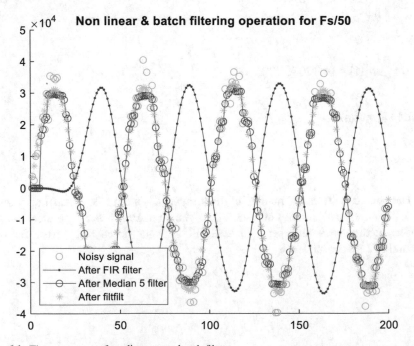

Fig. 6.1 Time response of median versus batch filters

6.4   Conclusions

In this chapter a non-linear median filter and a batch filter were described and demonstrated, the median filter has no advantage with respect to Gaussian noise attenuation but unlike a regular filter has a minor delay, when batch filtering is possible it has both an improved noise rejection and no delay, batch processes are usually done offline.

The median filter may have advantage with respect to noise attenuation for a different noise model.

Chapter 7
Interpolators

Abstract In this chapter we explain how to perform interpolation and fractional interpolation / decimation, and ways to implement them, a unique polyphase implementation is presented.

Keywords Interpolator · Filter · Fractional · Polyphase · interp1 · Resample

7.1 Introduction

In this chapter we explain how to perform interpolation and fractional interpolation / decimation, and ways to implement them.

7.2 Interpolator Design

An interpolator resamples the input data by R times the original sampling frequency followed by a LPF, or in MATLAB terms

[ymat, B] = interp(xin,R);

where ymat is the interpolated output, B is the interpolation filter performed at R*Fs, and xin is an N length array sampled at Fs.

Interpolation is first built by inserting R-1 zeros between any 2 original samples as follows:

```
xout = zeros(1,R*N);
for k = 1:N
  xout(R*(k – 1) + 1) = R*xin(k);
end
```

© The Author(s), under exclusive license to Springer Nature Switzerland AG 2022
A. Dickman, *Verified Signal Processing Algorithms in MATLAB and C*,
https://doi.org/10.1007/978-3-030-93363-0_7

If we perform an FFT with sampling frequency of R*Fs for Fs = 200 KHz and R = 16, we get (Fig. 7.1)

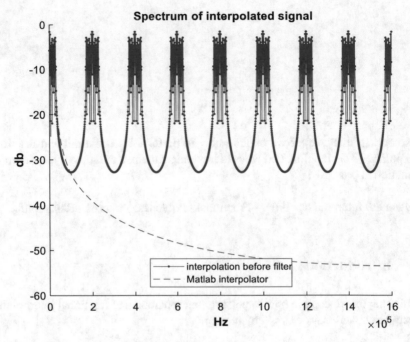

Fig. 7.1 Frequency response of the interpolation operation

xout was multiplied by R to compensate for the power loss caused by inserting R-1 zeros between any 2 samples of xin.

It is also possible to use zero order hold (ZOH) interpolation as follows:

```
xout = zeros(1,R*N);
ind = 1;
for k = 1:R:N
  xout(k : (k + R − 1)) = xin(ind);
  ind = ind + 1;
end
```

The spectrum of the interpolated signal includes R-1 images of the original spectrum from each side of 0 Hz, and the role of the interpolation filter is to remove them, if we design our own interpolation filter and present its response compared to the MATLAB filter it looks as

Code 7.1

```
Fs = 200e3;
Fin = 1000;
N = 4096;
t = (0:N-1)/Fs;

% Original signal
xin = 0;
for k = 1:30
  xin = xin + sin(2*pi*k*Fin*t + rand(1));
end

F1 = R*Fs/2/2048*(0:2047);

% MATLAB interpolation filter
B = B/(sum(B) + eps);
lenB = length(B)
hb = freqz(B,1,F1,R*Fs);

% Our interpolation filter
dm = fdesign.lowpass('Fp,Fst,Ap,Ast',30e3,Fs/2,0.75,74,R*Fs);
hm = design(dm);
b13 =  hm.numerator;
b13 = b13/(sum(b13) + eps);
lenb13 = length(b13);
hb13 = freqz(b13,1,F1,R*Fs);
```

As seen from the following response, the Fp frequency of our filter is set to be the edge of the input signal spectrum and Fst is set to Fs/2 Hz as the original input frequency cannot exceed Fs/2, both filters render 129 coefficients but our filter looks better (Fig. 7.2).

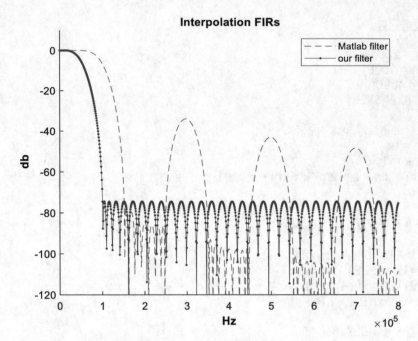

Fig. 7.2 Frequency response MATLAB versus our suggested filter

The processing time result on an I5 2.9G CPU is 211 nS per sample in floating point format using code 3.1.6 in Sect. 3.1.4, performed on R*Fs sampling frequency and decimation 1.

7.2.1 Fractional Interpolation/Decimation

It is often required to resample the input data by P / Q times the original sampling frequency using a LPF, where P and Q are natural numbers or in MATLAB code

[ymat, B1] = resample(xin,P,Q);

where ymat is the resampled output, B1 is the interpolation filter performed at P*Fs, and xin is an N length array sampled at Fs, when P < Q the final sampling frequency is lower than Fs.

First it is recommended to verify that the P,Q pair are minimal by using the MATLAB function

[P, Q] = rat(P / Q);

The straightforward solution would be to first interpolate by P and then decimate by Q, when P,Q are relatively small this simple solution might be chosen.

When the final sampling frequency is less than twice the highest frequency contained in xin, B1 should pass only ½ of the final sampling frequency if the signal is real and the final frequency if the signal is complex. The decimation part does not need a filter but only dilution by Q.

Since only 1:Q:N*P outputs are required it is a waste to compute all N*P outputs, an efficient algorithm would be to use a polyphase filter.

7.2.1.1 Polyphase Filter Design

A polyphase FIR filter is computed at the original sampling rate, making only the required outputs, the algorithm highlights are as follows:

a. Design an interpolation filter b13 at P*Fs Hz as in code 7.1 above, where Fp is the edge of the input spectrum, the filter length n should be a whole multiple of P, after a first design and having a preliminary number of coefficients, the requirements Ap, Ast are slightly stiffened till the closest whole multiple of P is achieved from above.
b. Prepare P sets from the filter, each set is b13(i:P:end), i = 1, ..., P.

Assume that we have a signal sampled at 10 Mhz and wish to resample it by 19/13, a MATLAB code that does it is

Code 7.2

```
fs = 10e6;
N = 1e4;
t = (0:N-1)/fs;
f1 = 10000;
fc = 2e6;
P = 19;
Q = 13;
Fs = fs*P;   % Interpolation

% Interpolation filter design
dm = fdesign.lowpass('Fp,Fst,Ap,Ast',4e6,fs/2,.4,82,Fs); % (570)
hm = design(dm);
b13 =  hm.Numerator;
b13 = b13/(sum(b13) + eps);
n = length(b13);

F1 = Fs/2/8192*(0:8191);
[H,W] = freqz(b13,1,8192);

figure (112), hold on
plot(W*Fs/2/pi,20*log10(abs(H)),'b.-')
title ('Polyphase filter','fontweight','b')
xlabel('Hz','fontweight','b')
ylabel('db','fontweight','b')
axis([0 Fs/8 -120 10])
hold off

%
% Reference
%
sig = 10000*(.3 + (1 + 0.5*(sin(2*pi*f1*t)) – 0.25*(sin(2*pi*5*f1*t))).*cos(2*pi*fc*t));
sigi = zeros(P*length(sig),1)';   % upsampling
sigi(1:P:end) = sig;
sigf = filter(b13,1,sigi);
Ref = sigf(n:Q:end);                % downsampling

%
% Polyphase filter implementation
%
c1 = n/P;
len = c1;
c2 = P;
slice = sig(c1:-1 : (c1 – (len – 1)));
bb = b13(c2:P:end);

for q = 1:floor(N*P/Q – n/Q)

  res = sum(bb.*slice);
```

```
if (abs(res – Ref(q)) > 1e-11)
  display('alg error');
end

c1 = c1 + floor(Q/P);
c2 = c2 + rem(Q,P);

if (c2 > P)
  c2 = c2 – P;
  c1 = c1 + 1;
end

slice = sig(c1:-1 : (c1 – (len – 1)));
bb = b13(c2:P:end);

end
```

The designed interpolation filter response is as follows.

The filter in Fig. 7.3 has 570 coefficients, from which 19 sets of 30 coefficients are derived, if the filter's ripple or stopband attenuation should be improved, then addition of whole multiples of 19 is required.

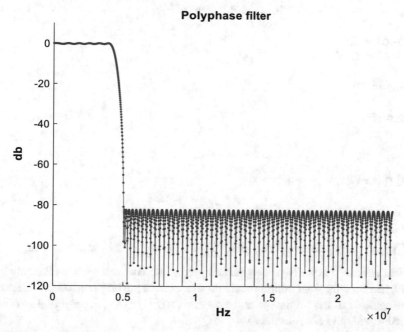

Fig. 7.3 Interpolation filter for resampling by 19/13

F_{st} is set to ½ of the original fs as the input signal cannot exceed it in order to keep the Nyquist theorem, and F_p is set to 80% of it, if F_p would be closer to F_{st} that would increase the number of the filter's coefficients.

The test signal sig undergoes the full fractional interpolation procedure and then compared to the output of the algorithm, the error message "alg_error" should not appear for any output.

For implementing MATLAB code 7.2 in C the filter's coefficients are stored in an *.h and poly_len is 570, floating point format is used, the code is

Code 7.3

```
void polyphase(float *x, int P, int Q, int n, float *y)
{
int k, m, len, c1, c2, num_outputs;
float sum1;

c1 = poly_len/P – 1;
len = c1 + 1;
c2 = P – 1;
num_outputs = n*P/Q – poly_len/Q;
for (k = 0; k < num_outputs; k++)
  {

  sum1 = (float)0.;
  for (m = 0; m < len; m++)
   sum1 += (x[c1 – m] * poly_coeff[c2 + m*P]);

  c1 = c1 + Q/P;
  c2 = c2 + Q % P;

  if (c2 >= P)
   {
   c2 = c2 – P;
   c1 = c1 + 1;
   }

  y[k] = sum1;
  }
}
```

The processing time result on an I5 2.9G CPU is 121 nS per output sample.

Next assume that we have a signal sampled at 10 Mhz and wish to resample it by 23/47, this time the final sampling rate is smaller than the original 10 Mhz, therefore Fst is again ½ of the original fs and Fp is fs*P/(2*Q) as the input signal fc cannot exceed it without being aliased to (fs*P/Q – fc).

Code 3.1.1 at Sect. 3.1.2 should be useful to check the result of the aliased frequency, a MATLAB code to implement the above example is as follows:

Code 7.4

```
fs = 10e6;
N = 1e4;
t = (0:N-1)/fs;
f1 = 10000;
fc = 1e6;
P = 23;
Q = 47;
Fs = fs*P;

% The reference signal
sig = 10000*(.3 + (1 + 0.5*(sin(2*pi*f1*t)) −
0.25*(sin(2*pi*5*f1*t))).*cos(2*pi*fc*t));

% Interpolation filter design
dm = fdesign.lowpass('Fp,Fst,Ap,Ast', fs*P/(2*Q),fs/2,0.22, 82,Fs); % (299)
hm = design(dm);
b13 =  hm.Numerator;
b13 = b13/(sum(b13) + eps);
n = length(b13);
```

The rest of code 7.4 is identical to code 7.2, the filter has 299 coefficients, from which 13 sets of 23 coefficients are derived, the interpolation filter response is as follows.

If the ripple or stopband attenuation of the filter in Fig. 7.4 should be improved, then addition of whole multiples of 23 is required.

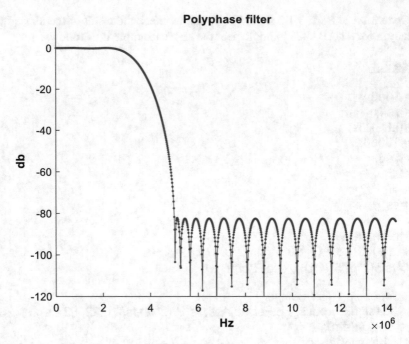

Fig. 7.4 Interpolation filter for resampling by 23/47

The C implementation is identical to code 7.3, the processing time result on an I5 2.9G CPU is 76 nS per sample in floating point format.

7.2.1.2   Fractional Interpolation/Decimation Using MATLAB interp1 Method

This method is simpler than the polyphase filter and implements the algorithm of the interp1 MATLAB function in linear mode, first xin is interpolated by a natural number M that is the closest to P/Q, then a P1/Q1 linear interpolation is performed so that M*P1/Q1 = P/Q, for example, if P/Q = 157/85, then interpolate by M = 2 and use P1/Q1 = 157/170, and if P/Q = 19/9, then M = 2 and P1/Q1 = 19/18, a MATLAB code that does that is

<u>Code 7.5</u>

```
Fs = 8e5;
N = 20000;
t = (0:N-1)/Fs;
Fin = 0.125*Fs;
R = 2;
Inp1 = 1000*sin(2*pi*Fin*t);
inp = interp(inp1,R);

P1 = 19;
Q1 = 18;

adv = Q1/P1;
acc = 0;

for k = 1:round(R*N*P1/Q1) – 2)

  k1 = floor(acc) + 1;
  del = acc – (k1 -1);
  y2(k) = inp(k1) + del*(inp(k1 + 1) – inp(k1));
  acc = acc + adv;

end

% The true interpolated samples
t2 = (0:length(y2)-1)/Fs/R/P1*Q1;
inpt = 1000*sin(2*pi*Fin*t2);

% The MATLAB resample operation
y1 = resample(inp,P1,Q1);

% Error calculation of 'resample' versus true samples
y1 = y1(1:length(inpt));
err = y1 – inpt;
err_db_resample = db(std(err)/std(inpt));

% Error calculation of the algorithm versus true samples
err = y2 – inpt;
err_db_interp1 = db(std(err)/std(inpt));
```

err_db expresses the difference in db between the algorithm output and the true samples values or the output of the MATLAB function resample relative to the true samples values, it does not depend on the ratio between P1 and Q1 but on the ratio between Fs and Fin, the higher the ratio the err_db is better (lower). If it should be improved, then R should be increased (provided that there is no problem to compute the interpolation filter at higher frequency) and then correct P1,Q1 so that the final fractional interpolation / decimation is maintained.

For code 7.5 err_db_resample is -59 db which shows that the resample function is accurate, err_db_interp1 is 22 db higher but is rather simple to implement, if we check the relative error of the interp1() MATLAB function using

```
t = (0:N-1)/Fs;
inp1 = 1000*sin(2*pi*Fin*t);
t2 = (0:length(y2)-1)/Fs/R/P1*Q1;
interp_mat = interp1(t,inp1,t2,'linear');
```

We get -25 db relative error which is 12 db worse than code 7.5 which implements the same operation, if we use mode "spline" as follows:

```
interp_mat = interp1(t,inp1,t2,'spline');
```

We get relative error of -62 db which is even better than "resample" and the most accurate mode out of the 8 interp1 modes, but also the most complicated to implement and the slowest one, it is slower by about 22 times relative to "linear" mode according to tic-toc measurement.

A C code that implements code 7.5 is

Code 7.6

```
void interp_fine(float *x, int P, int Q, int n, float *y)
{
int k, k1, len;
double adv, acc, del;

len = n*P/Q + 1;
adv = (double)Q / (double)P;
acc = (double)0.;

for (k = 0; k < len; k++)
  {
  k1 = (int)acc;
  del = acc - (double)k1;
  y[k] = x[k1] + (float)(del * (double)((x[k1 + 1]  - x[k1])));
  acc += adv;
  }
}
```

The processing time result on an I5 2.9G CPU is 33 nS per sample in floating point format after the interpolation by 2.

7.3 Conclusions

In this chapter sampling frequency interpolation design was presented including fractional interpolation / decimation. 2 methods for fractional interpolation / decimation were presented, the processing time for the interp1 method in linear mode is faster but this method may suffer from degraded SNR.

Chapter 8
Direct Digital Synthesis

Abstract In this chapter we explain how to design and implement digital direct synthesizers (DDS), how to modulate their frequency or amplitude and demodulate, respectively, for verification.

Keywords DDS · FM · Modulation · Demodulation · AM · Receiver

8.1 Introduction

In this chapter we explain how to design and implement a digital direct synthesizer (DDS), how to modulate its frequency or amplitude and demodulate, respectively, for verification.

8.2 Direct Digital Synthesizer Details

A direct digital synthesizer (DDS) generates a sine wave, making the operation for real output in MATLAB for Fs/32 frequency as follows:

Code 8.1

```
N = 1e4;
Fs = 10.24e6;
f1 = Fs/32;
t = (0:N-1)/Fs;
dds_out = sin(2*pi*f1*t);
```

Like the NCO at Sect. 3.1.2 the DDS uses a sin table from 0 to 2*pi with 2^15 entries, described by a MATLAB code as

© The Author(s), under exclusive license to Springer Nature Switzerland AG 2022   107
A. Dickman, *Verified Signal Processing Algorithms in MATLAB and C*,
https://doi.org/10.1007/978-3-030-93363-0_8

```
ScaleTab = 15;
SinTable = round(sin([0:2^ScaleTab -1)]/(2^ScaleTab)*2*pi)*(2^ScaleTab – 1));
```

The table coefficients are written to an *.h file as described in Sect. 2.2.2, ScaleTab is chosen by the required resolution of the generated frequency, which is the sampling frequency divided by the table size 2^ScaleTab, and by verifying that spurs and noise at the output are tolerable.

A MATLAB code that implements DDS operation is

Code 8.2

```
acc = 0;
Fc = round(f1/Fs*2^ScaleTab);
sin_gen = zeros(1,N);
for k = 1:N

  sin_adr = acc;
  if (sin_adr == 0)
    sin_adr = 1;
  end

  sin_gen(k) = SinTable(sin_adr);
  acc = mod(acc + Fc, 2^ScaleTab);

end
```

The phase increment dp is round(f1/Fs*2^ScaleTab), the software produces a short array of the outputs, if f1 is negative, then dp is round((Fs + f1)/Fs*2^ScaleTab), a C code that implements code 8.2 is

Code 8.3

```
void dds(int n, int dp, short *y)
{
int acc, i;

acc = 0;
for (i = 0; i < n; i++)
  {
  y[i] = Sin_Tab[acc];
  acc = (acc + dp) & 0x7fff;
  }
}
```

Code 8.3 consumes 28 nS per sample on an I5 2.9 GHz.

A common requirement is to modulate the DDS output, we will focus on FM and AM modulation.

8.2.1   *Frequency Modulation of a Direct Digital Synthesizer*

Frequency modulation (FM) changes the frequency of a DDS by a modulating signal which is characterized by the audio frequency Faudio and the rate at which the frequency is changed Fdev, a MATLAB code that implements such a modulation is

Code 8.4

```
Fs = 10.24e6;
N = 2^12;
t = (0:N-1)/Fs;
Fin = 1e6;
quan = 15;
scale = 64;
j = sqrt(-1);
Faudio = 5000;
Fdev = 40000;

SinTable = round(sin([0:2^quan -1])/(2^quan)*2*pi)*(2^quan – 1));

Mod_FM = Fdev*sin(2*pi*Faudio*t);
Mod_VCO_FM = Mod_FM*2^quan;

dp = round(Fin/Fs*2^quan);
dp_mod = round(Faudio/Fs*2^quan);

acc1_FM = -dp;
for k = 1:N

 acc1_FM = mod(acc1_FM + dp + Mod_VCO_FM(k)/Fs,2^quan);
 ind1_FM = mod(round(acc1_FM),2^quan);
 if (ind1_FM == 0)
  ind1_FM = 1;
 end
 Xdds_FM(k) = scale* SinTable(ind1_FM);

end
```

This code is implemented in C using a 2^15 elements sin table as in code 8.1, the C code is as follows:

Code 8.5

```
void dds_fm(int n, int Faudio, int Fdev, int Fin, int Fs, int scale, int *y)
{
int acc_mod, i, dp_mod, dp, del_mod, Fsh, ind;
double acc, kf, del2;

dp_mod = Faudio * 32768 / Fs;
Fsh = Fs >> 1;
if (((Faudio * 32768) % Fs) >= Fsh)
  dp_mod += 1;
dp = (int)((double)Fin * (double)32768. / (double)Fs + (double)0.5);

kf = (double)Fdev/(double)Fs;
acc = - (double)dp;
acc_mod = -dp_mod;

for (i = 0; i < n; i++)
  {
  acc_mod = (acc_mod + dp_mod) & 0x7fff;

  del_mod = Sin_Tab[acc_mod];
  del2 = kf * (double)del_mod;

  acc = acc + (double)dp + del2;

  if (acc >= (double)32768.)
    acc -= (double)32768.;

  ind = (int)(acc + (double)0.5);
  y[i] = scale*Sin_Tab[ind];

  }
}
```

If Fin is negative, then Fin is replaced by (Fs + Fin), the processing time of code 8.5 is 38 nS per sample on an I5 2.9 GHz.

The spectrum of the MATLAB and C code output are as follows (Fig. 8.1):

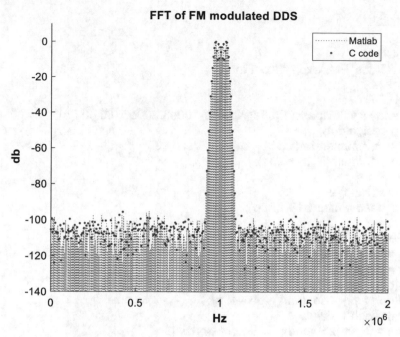

Fig. 8.1 Frequency response of an FM modulated DDS

The std of the difference between the outputs of the MATLAB and C code relative to the output std is -74 db.

In order to verify that the DDS was correctly modulated we describe a suitable receiver and demodulator as follows.

8.2.2 Receiver—Frequency Demodulator for Verification

A receiver—frequency demodulator in MATLAB to verify the FM modulation is as follows:

Code 8.6

```
% NCO
nco = Xdds_FM.*exp(-j*2*pi*Fin*t);

% LP FIR design
dm = fdesign.lowpass('Fp,Fst,Ap,Ast',100e3,200e3,0.5,80,Fs);
hm = design(dm);
b13 =  hm.numerator;
b13 = b13/(sum(b13) + eps);

% Reject image
base_band = filter(b13,1,nco);

% FM demodulator
ang = angle(base_band);
dph = diff(ang);
dph(dph < -pi) = dph(dph < -pi) + 2*pi;
dph(dph > pi) = dph(dph > pi) – 2*pi;
fff_fm = dph*Fs/2/pi;
fff_fm = fff_fm(250:end);   % Ignore transient
```

The modulated and demodulated outputs are shown as follows (Fig. 8.2):

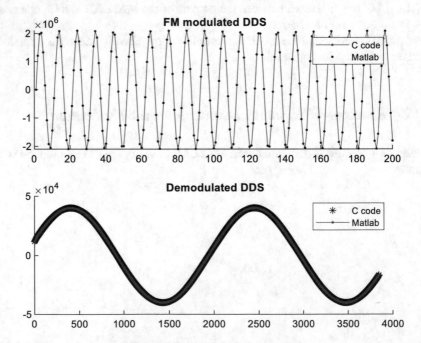

Fig. 8.2 Time response of an FM modulated and demodulated DDS

8.2.3   *Amplitude Modulation of a Direct Digital Synthesizer*

Amplitude modulation changes the amplitude of a DDS by a modulating signal which is characterized by the audio frequency Faudio, the depth of modulation is the modulation index with values of 0 to 1, a MATLAB code that implements such a modulation is

Code 8.7

```
Fs = 1.024e6;
N = 2^13;
t = (0:N-1)/Fs;
Fin = 1e5;
quan = 15;
j = sqrt(-1);
Faudio = 10000;
mod_index = 0.5;

SinTable = round(sin([0:2^quan -1)]/(2^quan)*2*pi)*(2^quan – 1));

Mod_VCO_AM = 1 + mod_index*sin(2*pi*Faudio*t);

dp = round(Fin/Fs*2^quan);
dp_mod = round(Faudio/Fs*2^quan);
acc1_AM = -dp;

for k = 1:N

  acc1_AM = mod(acc1_AM + dp,2^quan);
  ind1_AM = acc1_AM;
  if (ind1_AM == 0)
    ind1_AM = 1;
  end
  Xdds_AM(k) = SinTable(ind1_AM)*Mod_VCO_AM(k);

end
```

This code is implemented in C using a 2^15 elements sin table as in code 8.1, the C code is as follows:

Code 8.8

```
void dds_am(int n, int Faudio, float index, int Fin, int Fs, int *y)
{
int acc, acc_mod, i, dp_mod, dp, del_mod, Fsh;
float del2;

dp_mod = Faudio * 32768 / Fs;
Fsh = Fs >> 1;
if (((Faudio * 32768) % Fs) >= Fsh)
  dp_mod += 1;
dp = (int)((double)Fin * (double)32768. / (double)Fs + (double)0.5);

acc = -dp;
acc_mod = -dp_mod;


for (i = 0; i < n; i++)
  {
  acc_mod = (acc_mod + dp_mod) & 0x7fff;

  del_mod = Sin_Tab[acc_mod];
  del2 = (index * (float)del_mod) / (float)32768.;

  acc = (acc + dp) & 0x7fff;

  y[i] = (int)(((float)1. + del2)* (float)Sin_Tab[acc]);
  }
}
```

If Fin is negative, then Fin is replaced by (Fs + Fin), the processing time of code 8.8 is 35 nS per sample on an I5 2.9 GHz.

The spectrum of the MATLAB and C code output are as follows (Fig. 8.3):

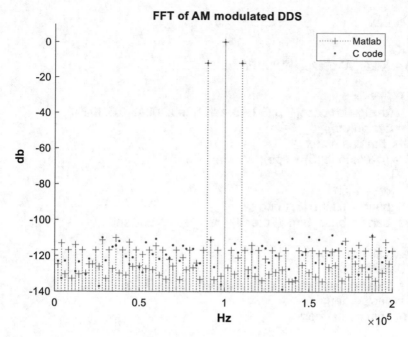

Fig. 8.3 Frequency response of an AM modulated DDS

The std of the difference between the outputs of the MATLAB and C code relative to the output std is -74 db.

In order to verify that the DDS was correctly modulated we describe a suitable receiver and demodulator as follows.

8.2.4 Receiver—Amplitude Demodulator for Verification

A receiver—AM demodulator in MATLAB to verify the AM modulation is as follows:

Code 8.9

```
% NCO
nco = Xdds_AM.*exp(-j*2*pi*Fin*t);

% LP FIR design
dm = fdesign.lowpass('Fp,Fst,Ap,Ast',50e3,100e3,0.5,80,Fs);
hm = design(dm);
b13 =  hm.numerator;
b13 = b13/(sum(b13) + eps);

% Reject image
base_band = filter(b13,1,nco);
base_band = base_band(70:end);  % Ignore transient

% AM demodulator
% HP FIR design
dm = fdesign.highpass('Fst,Fp,Ast,Ap',1000,4000,60,1,Fs);
hm = design(dm);
b23 =  hm.numerator;

P = abs(base_band);
fff_am = filter(b23,1,P);
fff_am = fff_am(800:end);   % Ignore transient
```

The modulated and demodulated outputs are shown as follows (Fig. 8.4):

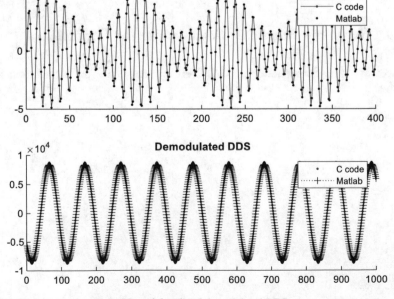

Fig. 8.4 Time response of an AM modulated and demodulated DDS

8.3   Conclusions

In this chapter we explained how to design and implement a digital direct synthe-sizer (DDS), how to modulate its frequency or amplitude and demodulate, respec-tively, for verification.

Chapter 9
Inverse Fast Fourier Transform

Abstract In this chapter we explain the inverse fast Fourier transform (IFFT), how to implement IFFT by using FFT, and how to modulate all bins.

Keywords IFFT · FFT · Peak to average · Modulation

9.1 Introduction

In this chapter we will explain the inverse fast Fourier transform (IFFT), how to implement IFFT by using FFT, and how to modulate all bins.

9.2 Inverse Fast Fourier Transform Details

IFFT (Inverse fast Fourier transform) is the opposite operation to FFT that renders the time response of a signal given its complex spectrum.

MATLAB expression for IFFT is

$$y = N*ifft(x,N)$$

which is an algorithm to compute faster than the IDFT (inverse discrete Fourier transform) expressed by

```
j = sqrt(-1);
for i = 1:N
  y(i) = N*sum(x.*exp(j*2*pi*(i – 1)/N*(0:N-1)));
end
```

where x is a complex or real N points FFT array and y is a size N complex output array, 1:N/2 input elements contain the positive frequencies $(0:N/2 - 1)*Fs/N$ and $(N+1):N$ input elements contain the negative frequencies $(-N/2 : -1)*Fs/N$.

The sequence of exponents $\exp(j*2*pi/N*(0 : N-1))$ is called twiddle factors as for FFT.

Like FFT, The IFFT computes the time response in $\sim N*\log_2 N$ operations for radix 2 algorithm and $\sim N*\log_4 N$ operations for radix 4 algorithm instead of N^2 operations that are required when calculating explicitly the IDFT.

An example to a typical MATLAB calculation of an ifft for a 10 bins input at 4.096 M samples per sec at a distance of 50,000 Hz apart from each other is

Code 9.1

```
Fs = 4.096e6;
N = 2^13;
i1 = 200;

inp = zeros(1,N);
inp(i1:100 : (i1 + 900)) = 1000;

Ifft = ifft(inp,N);
```

The time response of the amplitude is as follows (Fig. 9.1).

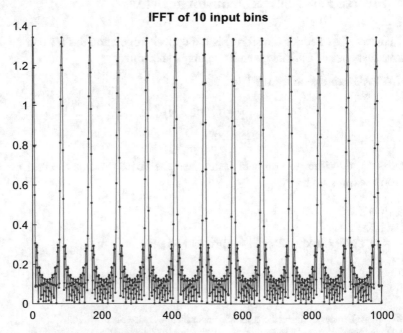

Fig. 9.1 Time response of a multi bin IFFT

No phases were attached to the non-zero bins, and as we can see the output contains significant peaks, if we compute the peak to average ratio for the ifft output by using the formula max(abs(Ifft))/std(Ifft) then when the distance between the bins approaches 1 then for equal amplitudes bins the ratio is ~sqrt(n1) where n1 is the number of non-zero bins.

This ratio may be minimized by attaching certain phases of the form $\exp(j*\phi_i)$ to each bin, an algorithm how to choose those phases is beyond the scope of this book.

9.3   Inverse Fast Fourier Transform Implementation

The IFFT is computed by $\log_2 N$ or $\log_4 N$ butterflies, since in practice each butterfly loses about ½ bit in resolution which decreases the dynamic range, then unless autoscaling is used, floating point implementation is preferred.

We offer 2 methods to implement IFFT using FFT that was described in detail at Sect. 3.2.4 (Table 9.1).

a. Use the property

IFFT(X) = conj(FFT(conj(X)))/N or in MATLAB code

Ifft_by_FFT = conj(fft(conj(inp),N)/N);

b. Use swapping [1] as follows in a MATLAB code

Ifft_by_FFT = fft(imag(inp) + j*real(inp),N)/N;
Ifft_by_FFT = imag(Ifft_by_FFT) + j*real(Ifft_by_FFT);

Table 9.1 The processing times for different sizes IFFT radix 2 on an I5 2.9G CPU

IFFT size	Processing time [µS]
128	7
256	16
512	33
1024	70
2048	160
4096	340
8192	700

and processing time for IFFT 4096 points radix 4 is 240 µS, those times exclude additional time for conj or swapping.

9.4 Frequency Modulation of Inverse Fast Fourier Transform Output

A MATLAB code for FM modulating all bins with the same modulation is as follows:

Code 9.2

```
Fs = 4.096e6;
N = 2^13;
j = sqrt(-1);
t = (0:N-1)/Fs;
i1 = 100;
quan = 15;
Faudio = 1000;
Fdev = 5000;

inp = zeros(1,N);
inp(i1:200 : (i1 + 1200)) = 1000;

Ifft = ifft(inp,N);

ModSig_FM = cumsum(Fdev*sin(2*pi*Faudio*t));
ModVco_FM = 2^quan*sin(2*pi*ModSig_FM/Fs);


for k = 1:N
  Xdds_FM(k) = real(Ifft(k))*ModVco_FM(k);
end
```

This code is implemented in C using a 2^{15} elements sin table as in code 8.1, the C code is as follows:

Code 9.3

```
void ifft_fm(int n, int Faudio, int Fdev, float *Ifft, int Fs, int *y)
{
int i, acc_FM, dp_mod, Fsh, del, ind;
double acc, kf, del2;

kf = (double)Fdev/(double)Fs;
dp_mod = Faudio * 32768 / Fs;
Fsh = Fs >> 1;
if (((Faudio * 32768) % Fs) >= Fsh)
  dp_mod += 1;

acc_FM = -dp_mod;
acc = (double)0.;
for (i = 0; i < n; i++)
  {
  acc_FM = (acc_FM + dp_mod) & 0x7fff;

  del = Sin_Tab[acc_FM];

  del2 = kf * (double)del;

  acc = acc + del2;

  if (acc >= (double)32768.)
    acc -= (double)32768.;

  if (acc < (double)0.)
    acc += (double)32768.;

  ind = (int)(acc + (double)0.5);

  y[i] = (int)(Ifft[i] * (float)Sin_Tab[ind]);
  }
}
```

The processing time of code 9.3 is 21 nS per sample on an I5 2.9 GHz.
    The spectrum of the MATLAB and C code output are as follows (Fig. 9.2):

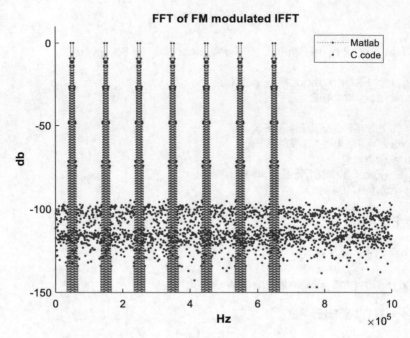

Fig. 9.2 Frequency response of a multi bin FM modulated IFFT

9.5   Conclusions

In this chapter we explained the IFFT transform and how to implement IFFT by using FFT, we showed how to modulate the frequency of all bins.

Reference

1. Texas instruments : ftp://ftp.ti.com/pub/tms320bbs/c67xfiles/cfftr2.asm, 1998

Chapter 10
Hilbert Transform

Abstract In this chapter we describe the Hilbert transform and its implementation.

Keywords Hilbert transform · FFT · IFFT · NCO · Filter

The Hilbert transform converts a real signal to complex, used when phase information of the signal is required, a MATLAB code to describe it is

<u>Code 10.1</u>

```
N = 2^14;
Fs = 10.24e6;
t = (0:N-1)/Fs;
Fin = 1e5;

x = 10000*sin(2*pi*Fin*t);
y1 = hilbert(x);
```

The transform performs an FFT, zeros the negative frequencies part and returns to time domain by IFFT, a detailed MATLAB implementation is

© The Author(s), under exclusive license to Springer Nature Switzerland AG 2022
A. Dickman, *Verified Signal Processing Algorithms in MATLAB and C*,
https://doi.org/10.1007/978-3-030-93363-0_10

Code 10.2

```
h1 = fft(x);
for i = 1:N

  if ((i == 1) | (i == (N/2 + 1)))
    h1(i) = h1(i);
  elseif ((i > 1) & (i <= N/2))
    h1(i) = 2*h1(i);
  else
    h1(i) = 0;
  end

end
y2 = ifft(h1);
```

The time response of the input x and complex output y2 is as follows (Fig. 10.1):

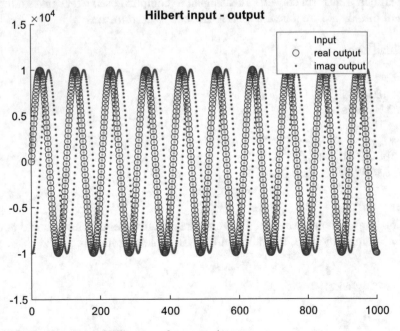

Fig. 10.1 Time response of Hilbert transform on a sine wave

The process of converting a real signal to complex may be achieved by using an NCO followed by a low pass filter, which is preferable in terms of processing time, since the Hilbert transform requires an FFT plus IFFT processes in addition to NCO and filter if frequency shift is required.

Only when no frequency shift is required it is worth to use the Hilbert transform, the operations in code 10.2 besides the FFT and IFFT are negligible.

Chapter 11
Channelizer

Abstract In this chapter we explain the channelizer basics, compare it to narrow band processing with respect to processing time, we then describe methods to implement a channelizer and analyze the results.

Keywords Channelizer · Narrow band · Stream · Chebwin · flattopwin · FFT · NCO · WOLA

11.1 Introduction

In this chapter the channelizer basics are explained, comparison to narrow band processing with respect to processing time is made, we then describe methods to implement a channelizer and analyze the results.

11.2 Channelizer Fundamentals

A channelizer is an algorithm to produce multiple streams of narrow band channels in parallel, which may later be demodulated to get audio or digital symbols information.

The basic process is to perform consecutive FFTs where each one has skipped in time with some overlap, a MATLAB code for a basic channelizer with no frequency shaping is as follows, producing a complex IF stream at sampling frequency Fs/skip for a 10mS FM modulated input

© The Author(s), under exclusive license to Springer Nature Switzerland AG 2022   127
A. Dickman, *Verified Signal Processing Algorithms in MATLAB and C*,
https://doi.org/10.1007/978-3-030-93363-0_11

Code 11.1

```
Fs = 40.96e6;
res = 5000;
j = sqrt(-1);

Tpro = 10e-3;
N1 = round(Fs*Tpro);

t = (0:N1-1)/Fs;
Faudio = 500;
Fdev = 2000;
off_freq = 1000*res;

% FM modulated signal
tempMod = Fdev*cumsum(sin(2*pi*Faudio*t));
inp = 1000*exp(j*2*pi*off_freq*t + j*2*pi*tempMod/Fs);

%
% Make channelizer
%
N = 8192;
ind = off_freq/res + 1;
out1 = [];
skip = 4096;
k = 1;
phase_cor = exp(j*pi);

while (k < (length(inp) - N))

  sig = inp(k : (k + N - 1));
  Fout = fft(sig,N)/N;

  if (mod(ind,2) == 0)
    phase_cor = phase_cor*exp(j*pi);
  else
    phase_cor = 1;
  end

  out1 = [out1  phase_cor*Fout(ind)];
  k = k + skip;
end
```

The complex stream out1 is sampled at Fs/skip (10 Khz), since the bandwidth of inp is about 2*(Faudio + Fdev) or 5 Khz for an FM signal with Fdev/Faudio $\geq$ 4 then 10 Khz sampling frequency is high enough to construct the audio information, if we demodulate out1 as follows:

Code 11.2

```
% FM demodulator
ang = angle(out1);
dph = diff(ang);
dph(dph < -pi) = dph(dph < -pi) + 2*pi;
dph(dph > pi) = dph(dph > pi) – 2*pi;
fmout = dph*Fs/2/pi/skip;
```

The time response of the demodulated out1 is as follows (Fig. 11.1):

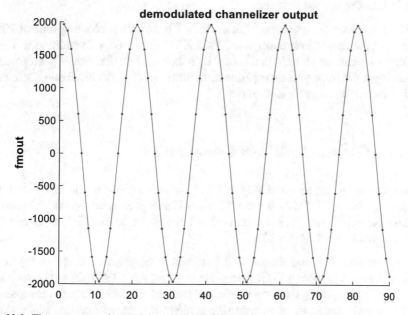

Fig. 11.1 Time response of an FM demodulated channel

which shows an amplitude of Fdev and frequency of Faudio as inserted, the term phase_cor is used for the odd channels since the skip is ½ of the FFT size, so that the phase_cor is multiplied by exp(j*pi) on each step or (1,-1,1,-1 …..1,-1).

11.2.1   Comparing Channelizer and Narrow Band Processing Times

Consider the example of code 11.1, with sampling frequency of 40.96 Mhz and check the processing time on an I5 2.9 GHz CPU for 1 sec sampling for both options, assume WOLA 4 window for the channelizer as will be explained in the next paragraph (Table 11.1).

Table 11.1 Comparison between processing times of channelizer and narrow band processes on an I5 2.9 GHz CPU

Channelizer operation	Processing time	Narrow band processing 1 channel	Processing time
10000 X WOLA 4	0.72 sec (per Sect. 3.2.2)	40.96e6 NCO operations	1.27 sec (per Sect. 3.1.2)
10000 X FFTs of 8192	7 sec (per Table 3.1 at Sect. 3.2.5.3)	4096 decimator	1.02 sec (per Sect. 3.1.6)
Total	7.72 sec		2.29 sec

From the above it is seen that even for 4 channels at all the channelizer is preferable in terms of processing times.

If we consider an eight times shorter FFT of 1024 with processing time of 70 µS each, then the channelizer consumes 80000 X 70 µS or 5.6 sec in addition to 80000 WOLA 4 operations of 1024 that do not take much, while the narrow band solution needs about the same processing time as before regarding the NCO and decimator, so the channelizer is preferable again.

11.2.2   Channelizer Design Considerations

When we design the required skip in input samples relative to the input sampling frequency and the FFT size, we should choose those parameters so that the spectral content of each channel will reside within 1 bin width or Fs / FFT_Size, and Fs / skip is about twice the bin width.

Another consideration should be the amount of neighboring channels rejection, since this is solved using a WOLA shaping window, the WOLA should be such that the sampling frequency Fs / skip is minimal but still channel separation is achieved. In the following table we summarize the amount of minimal rejection in db versus the WOLA order and window type used, following the details of Sect. 3.2.2

From Table 11.2 we deduce that using WOLA 3 to 5 windows are appropriate for the neighboring channels rejection and there is a tradeoff between a flat response at the passband zone and neighboring channels rejection, for the same WOLA order the Chebwin window rejects better the neighboring channels relative to flattopwin window.

Table 11.2 Neighboring channels rejection for different shaping windows

WOLA & window type	All neighboring channels rejection from 1 and up [db]
No window	13.5
Chebwin 106 no WOLA	3.4
Hanning no WOLA	6
Chebwin 106 WOLA 2	14.2
Chebwin 106 WOLA 3	35.5
Chebwin 106 WOLA 4	86
flattopwin WOLA 4	36
flattopwin WOLA 5	93.5

As the most processing time consumer in the channelizer is the FFT, a channelizer design using a smaller FFT and then NCO and filtering to isolate the subchannels contained within the initial FFT bandwidth is considered, let us call it a coarse / fine process.

Consider the above example where 10,000 FFTs of 8192 points and bin width of 5000 Hz are calculated in 1 sec, taking 7 s as per Table 3.1, if we add 10000 WOLA 4 calculations as well with 72 µS each, then 7.72 s are totally required.

If we perform 40000 FFTs of 2048 points instead at a skip of 1024 samples and bin width of 20,000 Hz in 1 s, then add the same number of WOLA 4 calculations, then (160 + 17) X 40000 consume 7.08 s.

If we need to produce 200 channels of 5000 Hz bandwidt h, then the NCO process consumes 200 X 31 nS X 40,000 = 248 mS, and from the 4 →1 decimator design in Sect. 3.1.5.2 we deduce that 200 X 0.6 mS = 120 mS are required for the respective decimator, this coarse / fine process takes totally 7.45 s.

A difference of 3.5% in processing times does not justify to use the coarse / fine process, which is more complicated, when the algorithms are implemented in firmware, the elevated use of hardware resources by using a four times bigger FFT is a consideration, but not for software considerations.

As the above example needs about eight times processing time than sampling time then if a multi core SBC is used, the solutions detailed in Sect. 3.2.6 may lower the processing to sampling times ratio.

In view of the above, a MATLAB code to implement a channelizer using WOLA 3 chebwin(106) window and FFT of 8192 is

Code 11.3

```
N = 3*8192;
ind = off_freq/res + 1;
out1 = [];
skip = 4096;
k = 1;
win_cheb3 = chebwin(N,106)';
win_cheb3 = win_cheb3/max(win_cheb3);
phase_cor = exp(j*pi);

while (k < (length(inp) – N))

  sig = inp(k : (k + N – 1));
  sigt = win_cheb3.* sig;
  sigt1 = sigt(1:N/3) + sigt(N/3 + (1:N/3)) + sigt(2*N/3 + (1:N/3));

  if (mod(ind,2) == 0)
    phase_cor = phase_cor*exp(j*pi);
  else
    phase_cor = 1;
  end

  Fout = fft(sigt1,N/3)/(N/3);
  out1 = [out1  phase_cor*Fout(ind)];

  k = k + skip;

end
```

If WOLA 4 flattopwin window and FFT of 8192 is used, then a respective MATLAB code to compute 4 adjacent channels is

Code 11.4

```
Fs = 40.96e6;
res = 5000;
j = sqrt(-1);
off_freq = 1000*res;
N = 4*8192;
ind = off_freq/res + 1;
out1 = [];
out2 = [];
out3 = [];
out4 = [];
skip = 4096;
k = 1;
win4 = flattopwin(N)';
win4 = win4/max(win4);
phase_cor1 = exp(j*pi);
phase_cor2 = 1;
t4 = (0:N-1)/Fs;
for k = 1:1000

  f1(k) = 50*k + off_freq - 16000;
  sig = sin(2*pi*f1(k)*t4 + rand(1));
  sigt = win4.* sig;

  sigt1 = sigt(1:N/4) + sigt(N/4 + (1:N/4)) + sigt(2*N/4 + (1:N/4))+...
    sigt(3*N/4 + (1:N/4));

  Fout = fft(sigt1,N/4)/(N/4);

  if (mod(ind,2) == 0)
    phase_cor1 = phase_cor1*exp(j*pi);
    phase_cor2 = 1;
  else
    phase_cor1 = 1;
    phase_cor2 = phase_cor2*exp(j*pi);
  end

  out1 = [out1 phase_cor1*Fout(ind)];
  out2 = [out2 phase_cor2*Fout(ind + 1)];
  out3 = [out3 phase_cor1*Fout(ind + 2)];
  out4 = [out4 phase_cor2*Fout(ind + 3)];

  k = k + skip;

end
```

The response of the four adjacent channels is as follows (Fig. 11.2):

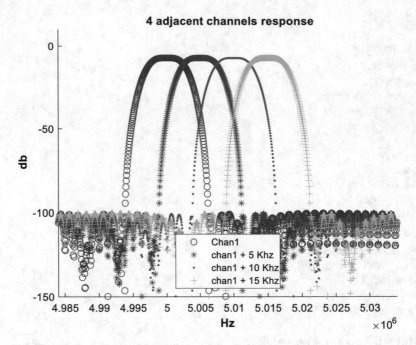

Fig. 11.2 Frequency response of four adjacent channels

It is possible to use a "small" FIR of 8–10 coefficients that will operate at Fs/skip sampling rate on every produced channel and will improve the neighboring channels rejection if a too low WOLA order is used, the designer should check if this 16–20 nS addition per sample per channel is preferable in comparison to increasing the WOLA order.

11.3 Conclusions

In this chapter we explained the channelizer basics, compared it to narrow band processing with respect to processing time and presented several implementation methods, we found that for software implementation the direct skipped FFTs method is preferred, since considerations such as FPGA limited resources are not relevant, the designer should verify that the sampling frequency and channel bandwidth are such that the CPU can process consecutive FFTs skipped in time in real time, with the aid of the processing time data given here.

Chapter 12
Correlation

Abstract In this chapter we describe the correlation and its implementation, a frequency domain method to accelerate processing time is presented, Pattern recognition is demonstrated and implemented.

Keywords Correlation · Reference · Samples · Figure of merit · conj · Pattern recognition

12.1 Introduction

In this chapter the correlation and its implementation are described, a frequency domain method to accelerate the processing time will be presented. The use of correlation for pattern recognition is demonstrated on a certain template which appears repeatedly in every observation slice at a constant delay from its start.

12.2 Correlation Details

Correlation is the most powerful tool to detect a signal provided that we have knowledge about its model.

Correlation is made between a reference array and a samples array, where the samples array is longer than the reference, the correlation operation slides the reference along the samples and searches for a match, a MATLAB code to describe the operation is

© The Author(s), under exclusive license to Springer Nature Switzerland AG 2022
A. Dickman, *Verified Signal Processing Algorithms in MATLAB and C*,
https://doi.org/10.1007/978-3-030-93363-0_12

Code 12.1

```
Time_shift = 135;
j = sqrt(-1);

% Simulating reference
xx = [0 0 0 1 1 0 1 0 0 0 1 1 0 1 0 1];
xx1 = [0 1 1 0 1 0 1 0 1 1 0 0 0 1 1 1];
xx2 = [xx xx1];
xx3 = [0 1 0 1 1 1 0 0 1 1 1 1 0 0 1 1];
ref = [xx1 ~xx1 ~xx2 ~xx1 ~xx2 xx3];

ref(ref == 0) = -1 + j;
ref(ref == 1) = 1 - j;

% Simulating samples
samples = randi([0 1],1,16*length(ref));
samples(samples == 0) = -1 + j;
samples(samples == 1) = 1 - j;
samples(Time_shift:Time_shift - 1 + length(ref))) = ref;

% Correlation operation
[cor_comp, lags] = xcorr(samples,ref);
cor = abs(cor_comp).^2;
```

cor and Lags are arrays with length $2*M - 1$ where M is max(length(samples), length(ref)).

An explicit implementation of the correlation in MATLAB is

Code 12.2

```
for k = 1:length(samples) - length(ref) + 1

  correl(k) = 0;
  for m = 1:length(ref)
    correl(k) = correl(k) + samples(k - 1 + m)*conj(ref(m));
  end
  correlA(k) = correl(k)*conj(correl(k));

end
```

correlA has a length of (length(samples) - length(ref) + 1) and alignment of xcorr output with this implementation is

```
cor_dis = cor(length(samples) - 1 + (1:length(samples) - length(ref) + 1)));
```

The correlation power can reach $4*(length(ref)^2)$ for $[-1 + j]$, $[1 - j]$ antipodal representation of complex samples if samples and ref have perfect correlation at any sample along samples, longer correlation length enhances maximum above noise, making it easier to identify a match and its timing, a figure of merit for the correlation quality is max(correlA) / mean(correlA).

A C implementation of code 12.2 is

Code 12.3

```c
void correlation(int *samp, int *ref, int ncor, int nref, int *y)
{
int i,k,cr,ci;

for (i = 0; i < ncor; i++)
  {
  cr = 0;
  ci = 0;
  for (k = 0; k < nref; k++)
    {
    cr += (samp[2*(i + k)]*ref[2*k] + samp[2*(i + k) + 1]*ref[2*k + 1]);
    ci += (samp[2*(i + k) + 1]*ref[2*k] − samp[2*(i + k)]*ref[2*k + 1]);
    }
  y[i] = cr*cr + ci*ci;
  }
}
```

The arrays samp and ref are arranged as consecutive pairs (real(0), imag(0), real(1), imag(1), ….real(end), imag(end)).

The correlations array results are as follows (Fig. 12.1) followed by a processing times table for different sizes of samp and ref on an I5 2.9 GHz CPU (Table 12.1):

From Table 12.1 we deduce that each multiplication of the samples or reference length doubles the processing time.

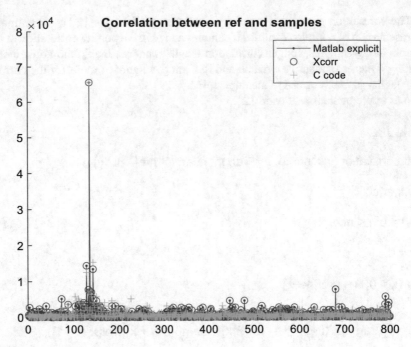

Fig. 12.1 Correlations array results

Table 12.1 Processing times for different sizes of samp and ref on an I5 2.9 GHz CPU

Samples length	Reference length	Processing time
2048	128	1600 µS
2048	64	830 µS
1024	128	760 µS
1024	64	410 µS

12.2.1   Fast Correlation by Using Fast Fourier Transform

A faster method to implement correlation uses the following property:

Correlation(X,Y) = IFFT(FFT(X).*FFT*(Y))

A more detailed description in MATLAB is as follows:

Code 12.4

```
Fref = fft([ref   zeros(1,length(samples) – length(ref))]);
Fsamples = fft(samples);

corF = ifft(Fsamples.*conj(Fref));
corF = abs(corF).^2;
```

The shorter array ref is extended to length(samples) by adding zeros, the fast correlation output with the regular correlation output are as follows, which shows identical results (Fig. 12.2)

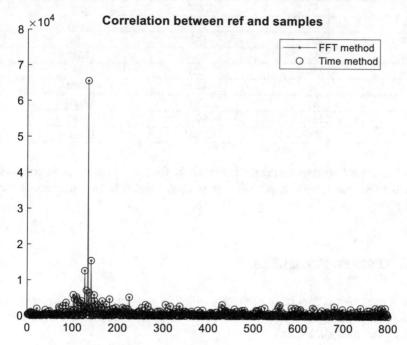

Fig. 12.2 Correlations array results using two methods

The operation X.*conj(Y) for complex interleaved arrays is implemented in C by the following code:

<u>Code 12.5</u>

```
void comp_array_mult(int *x1, int *x2, int n, int *y)
{
int i;

for (i = 0; i < n; i++)
  {
  y[2*i] = x1[2*i]*x2[2*i] – x1[2*i + 1]*x2[2*i + 1];
  y[2*i + 1] = x1[2*i + 1]*x2[2*i] + x1[2*i]*x2[2*i + 1];
  }
}
```

The processing time for this operation on an I5 2.9 GHz CPU is 7 nS per sample, taking account of the operations involved in the 2 methods and the data in Table 12.1 and Table 3.1 we have the following results for several cases (Table 12.2).

Table 12.2 Processing times for different sizes of samp and ref using two methods on an I5 2.9 GHz CPU

Samples length	Reference length	Regular method [µS]	Fast method [µS]
2048	64	830	500
2048	128	1600	500
2048	256	3200	500
4096	64	1600	1060
4096	128	3200	1060
4096	256	6400	1060
8192	64	3200	2180
8192	128	6400	2180
8192	256	12,800	2180

When the reference is relatively short there is about 40% reduction in processing time for the faster method, but when it is longer the reduction may reach more than 80%.

12.3   Pattern Recognition

Pattern recognition may be understood as finding a match between a certain template and a same segment length of samples, occurring at a specific sample relative to an observation slice start.

In the following example, the template array is hidden at a delay of Time_shift from observation_slice start which is analyzed for 50 such slices, an array of IIR correlators is used with length of observation_slice in samples, for each sample at which a template match is attained the respective correlation is charged, and discharged when there is no template match.

When the maximum correlation passes some predefined threshold then it is possible to use that sample index as the time point of the identification of the template.

The time segment used by each correlator is five observation slices, increasing it will improve the reliability of detection but also increase the transient time to reach the threshold.

A MATLAB code for the example is as follows:

Code 12.6

```
Time_shift = 100;
ref = [1 5 3 9 17];

% Prepare signal for identification
observation_slice = [randi([-20 20],1,Time_shift-1)
ref randi([-20 20],1,100*length(ref))];

% Prepare 50 observation_slice segments
samples = repmat(observation_slice,1,50);

correl = zeros(1,length(observation_slice));
len = 5;
kon = 2/len;
koff = 1 - kon;
index = 1;
max_cor = 0;
ind_cor = 0;
Thr = 0.5;

for k = 1:length(samples) - length(ref)

  if (length(intersect(ref,samples(k:(k + length(ref) - 1)))) == length(ref))
    correl(index) = koff*correl(index) + kon;
    correl_dis(k) = 0;
  else
    correl(index) = koff*correl(index);
    correl_dis(k) = correl(index);
  end


  if (correl(index) > max_cor)
    max_cor = correl(index);
    ind_cor = index;
  end
  max_correl(k) = max_cor;
  ind_correl(k) = ind_cor;

  index = index + 1;

  if (index > length(observation_slice))
    index = index - length(observation_slice);
  end

end
```

The samples, correlation values, and sample index of identification (same as Time_shift) are shown in the following figure (Fig. 12.3):

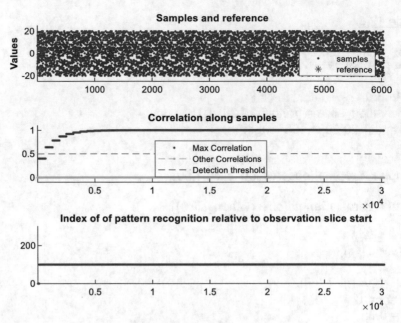

Fig. 12.3 Time response of pattern recognition products

An implementation of code 12.6 in C is as follows, the function receives a pointer to a chunk of samples and returns the sample index at which a given template is recognized and the value of the correlation at that point

Code 12.7

```c
float max_cor;
short ind_cor, index;
float kon = (float)0.4;
float koff = (float)0.6;
short cor_len = 604;
float correl[1000];

void pattern_rec(short *x, int n, short *ref, short len_ref, short num_slices,
short   *pIndex, float  *pCor_max, short init)
{
int i,k;
short c1;

if (init)
  {
  index = 0;

  max_cor = (float)0.;
  for (i = 0; i < cor_len; i++)
    correl[i] = 0;
  }
for (i = 0; i < (n - len_ref); i++)
  {

  c1 = 0;
  for (k = 0; k < len_ref; k++)
    {
    if (x[i + k] == ref[k])
      c1 += 1;
    }

  if (c1 == len_ref)
    correl[index] = koff*correl[index] + kon;
  else
    correl[index] = koff*correl[index];

  if (correl[index] > max_cor)
    {
    max_cor = correl[index];
    ind_cor = index;
    }
```

```
*pCor_max = max_cor;
*pIndex = ind_cor;

index += 1;

if (index >= cor_len)
  index -= cor_len;

}
}
```

The global variables correlation, index, max_cor are initialized only at the first time when the function is called, cor_len which is the observation slice length in samples should be known apriori.

The processing time for code 12.7 on an I5 2.9 GHz CPU is 18 nS per sample.

12.4   Conclusions

In this chapter the correlation and its implementation were described, a faster method to implement correlation by using FFT was shown.

When correlating a reference signal with samples that include a fingerprint of that reference, care should be taken to the length of the reference when the samples have a constant center of frequency.

If this center of frequency is not constant, then a too long correlation length would smear the correlation peaks and lower them.

If there is an unknown frequency change due to Doppler effect, for instance, then several references should be prepared, each having a certain frequency shift by multiplying by exp(j*2*pi*f/Fs*(0:N-1)) sample by sample where f id the shift and N the reference length, a two dimensional search should then be performed to find the correct frequency shift and then the reference fingerprint as well.

The use of correlation for pattern recognition was demonstrated on a certain template which appears every observation slice at a constant delay from its start.

Chapter 13
Adaptive Filters

Abstract In this chapter we describe adaptive filters and their implementation, reviewing different type of equalizers such as LMS, NLMS, RLS, and interference cancellation algorithms.

Keywords Adaptive filter · Equalizer · LMS · NLMS · Multi-input equalizer · Interference cancellation · RLS

13.1   Introduction

Adaptive filters are used whenever the environment is non-stationary, and interference cancellation, equalization, or echo cancellation are required, the filters are expected to follow the environment changes over time so that the mean square error between the followed variable and the estimated one is minimized.

Adaptive filter is also used for ISI (inter symbol interference) improvement.

In this chapter adaptive filters will be described, reviewing equalizers and interference cancellation algorithms.

13.2   Equalizers

An adaptive equalizer is an algorithm to estimate in real time a time varying response of a communication channel, enabling to receive better performance and throughput of that channel.

The inputs to an equalizer MATLAB description are as follows:

© The Author(s), under exclusive license to Springer Nature Switzerland AG 2022   145
A. Dickman, *Verified Signal Processing Algorithms in MATLAB and C*,
https://doi.org/10.1007/978-3-030-93363-0_13

```
N = 8192;
amp = 1;
Inp1 = amp*randn(N,1);   % Known training sequence

h = gaussfir(0.3);        % Simulated channel response
out1 = filter(h,1,inp1);  % Channel output

Eq_order = 16;
W = zeros(Eq_order,1);   % FIR Equalizer
```

The equalizer receives a known training sequence inp1 which should have a wide spectrum in order to enable a true channel estimation if it has a wide spectrum response too, the equalizer receives also the channel measured response to the above training sequence (out1), then an algorithm is performed that changes a predetermined order FIR W, so that the mean square error between out1 and inp1 filtered by W is minimized.

13.2.1   Least Mean Squares Algorithm

A Least Mean Squares (LMS) algorithm implementation in MATLAB is as follows [1]:

Code 13.1

```
Mu = 0.05;

for n = Eq_order:N

  D = inp1(n:-1 : (n – Eq_order + 1));
  est = sum(D.*W);
  e(n) = out1(n) – est;
  W = W + Mu*e(n)*D;

end
```

The algorithm convergence is controlled by the error e and the step size Mu, smaller Mu lengthens the convergence time and a too big Mu may destabilize the LMS algorithm.

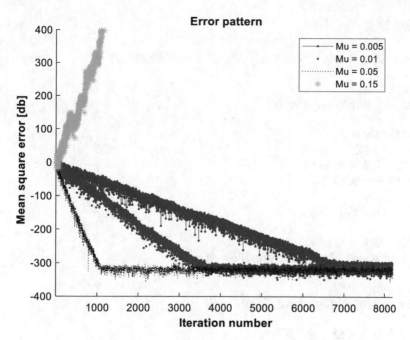

Fig. 13.1 LMS error pattern for different Mu values

The mean square error in db time behavior for different Mu values is (Fig. 13.1)

The convergence behavior for the LMS algorithm is inversely proportional to the variance of inp1 and to the length of W, so the following formula for Mu is used

Mu = KMu / Eq_order / var(inp1);

The KMu coefficient values for which the algorithm converges are 0.004 to 1.5, a code C implementation for code 13.1 is

Code 13.2

```
void lms(float *inp1, float *out1, int n, int neq, float *W)
{
int k, m;
float Mu, sum1, var, est, err;

sum1 = (float)0.;
var = (float)0.;
for (k = 0; k < n; k++)
  sum1 += inp1[k];

sum1 = sum1/(float)n;

for (k = 0; k < n; k++)
  var += ((inp1[k] – sum1)*(inp1[k] – sum1));

var = var/(float)n;
Mu = (float)0.5/ ((float)neq*var);

for (k = 0; k < neq; k++)
  W[k] = (float)0.;

for (k = (neq – 1); k < n; k++)
  {
  est = (float)0.;
  for (m = 0; m < neq; m++)
    est += (inp1[k – m] * W[m]);

  err = out1[k] – est;

  for (m = 0; m < neq; m++)
    W[m] = W[m] + Mu*err*inp1[k – m];
  }
}
```

   The processing time for a 16 elements W on an I5 2.9 GHz CPU is 101 nS per iteration, and 416 nS per iteration for a 64 elements W, the processing time is doubled for doubling the input length or equalizer order.

   In the following (Fig. 13.2) we demonstrate the equalizer operation on 3 examples of channel responses, the steady state coefficients for a FIR type channel response and two types of IIR type channel response as

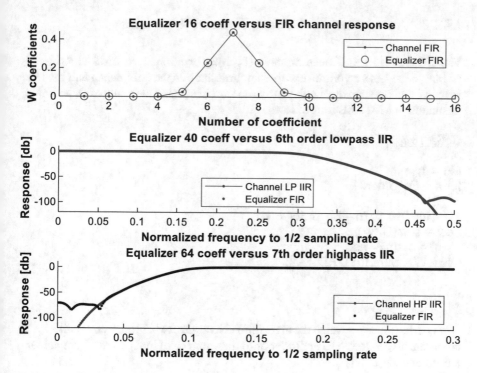

Fig. 13.2 Different LMS equalizers products

```
h = gaussfir(0.3);
out1 = filter(h,1,inp1);

[b1,a1] = butter(6,0.6);
out1 = filter(b1,a1,inp1);

[b2,a2] = butter(7,0.2,'high');
out1 = filter(b2,a2,inp1);
```

13.2.2   Normalized Least Mean Squares Algorithm

The Normalized Least Mean Squares (NLMS) or normalized LMS algorithm is similar to the LMS algorithm except that it normalizes the error dependent member in the W computation by dividing it by the sum of squared elements of D, an implementation in MATLAB is as follows [1]:

Code 13.3

```
Mu = 1;
for n = Eq_order:N

  D = inp1(n:-1 : (n – Eq_order + 1));
  est = sum(D.*W);
  e(n) = out(n) – est;
  W = W + Mu*e(n)*D /(D' * D);

end
```

The Mu values for which the algorithm converges are 0.2 to 1.9 and unlike LMS are insensitive to the equalizer order or to the variance of inp1, a code C implementation for code 13.3 is

Code 13.4

```c
void nlms(float *inp1, float *out1, int n, int neq, float Mu, float *W)
{
int k, m;
float sum1, est, err;

for (k = 0; k < neq; k++)
  W[k] = (float)0.;

for (k = (neq - 1); k < n; k++)
  {

  est = (float)0.;
  sum1 = (float)0.;
  for (m = 0; m < neq; m++)
    {
    est += (inp1[k - m] * W[m]);
    sum1 += (inp1[k - m] * inp1[k - m]);
    }

 err = out1[k] - est;

 for (m = 0; m < neq; m++)
   W[m] = W[m] + Mu*err*inp1[k - m] / sum1;

  }
}
```

The demonstrated examples in Figure 13.2 render the same results as LMS for the NLMS algorithm, the processing time for a 16 elements W on an I5 2.9 GHz CPU is 186 nS per iteration, and 742 nS per iteration for a 64 elements W, so the processing time is doubled for doubling the input length or equalizer order and is about 80% more than the LMS algorithm.

13.2.3 Multi-input Equalizer

The channel output or the measurement may consist of several independent sources, each having a different response to a known input, a description in MATLAB of an NLMS algorithm with three parallel different length equalizers is as follows:

Code 13.5

```
N = 16384;

x1 = 10*randn(N,1);      % Known training sequence1
x2 = 100*randn(N,1);     % Known training sequence2
x3 = randn(N,1);         % Known training sequence3

h1 = 20*gaussfir(0.2);   % Channel 1 response
h2 = gaussfir(1);        % Channel 2 response
[b,a] = butter(8,0.25);  % Channel 3 response

% Total response
d = filter(h1,1,x1) + filter(h2,1,x2) + filter(b,a,x3);

Eq_order1 = 16;
Eq_order2 = 24;
Eq_order3 = 84;
W1 = zeros(Eq_order1,1);     % Equalizer 1
W2 = zeros(Eq_order2,1);     % Equalizer 2
W3 = zeros(Eq_order3,1);     % Equalizer 3

mm = max([Eq_order1 Eq_order2 Eq_order3]);

%
% NLMS adapdation
%
Mu = .5;
for n = mm : N

  D1 = x1(n : -1 : n – Eq_order1 + 1);
  D2 = x2(n : -1 : n – Eq_order2 + 1);
  D3 = x3(n : -1 : n – Eq_order3 + 1);

  est = sum(D1.*W1) + sum(D2.*W2) + sum(D3.*W3);
  e(n) = d(n) – est;

  W1 = W1 + Mu*e(n)*D1/(D1' * D1);
  W2 = W2 + Mu*e(n)*D2/(D2' * D2);
  W3 = W3 + Mu*e(n)*D3/(D3' * D3);

end
```

The NLMS was chosen in order to ease normalization for several inputs, each using a different length equalizer, convergence is achieved for Mu values of 0.05 to 0.6 and insensitive to equalizers order or to the amplitudes of the 3 inputs.

The mean square error in db time behavior for Mu = 0.5 is (Fig. 13.3)

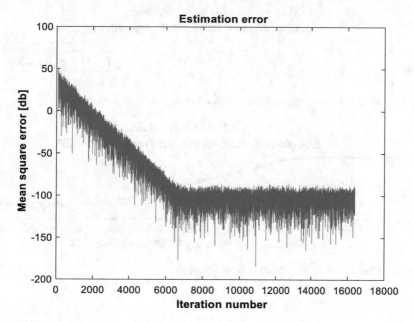

Fig. 13.3 Estimation error pattern for multi input equalizer

The three estimated steady state responses are as follows (Fig. 13.4):

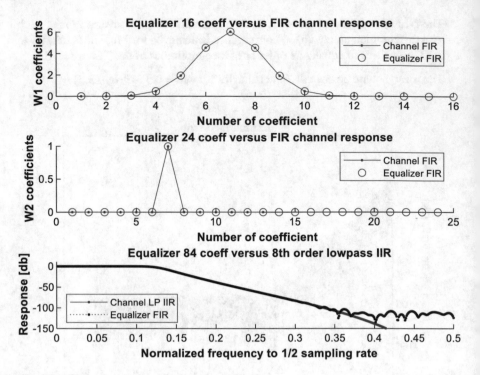

Fig. 13.4 Channels estimators for multi input equalizer

13.3   Interference Cancellation

This algorithm deals with a source which is disturbed by a known interference having a channel response and possibly a multi-path component with the presence of noise, the following MATLAB code describes an FM modulated source with such an interference, which may be heard by a PC sound device

<u>Code 13.6</u>

```
N = 32768;
j = sqrt(-1);
Fs = 32000;
t = (0:N-1)'/Fs;
offset = 20;
Faudio = 1000;
Fdev = 2000;

% FM modulated signal
tempMod = Fdev*cumsum(sin(2*pi*Faudio*t))/Fs;
inp1 = 10000*exp(j*2*pi*tempMod);

% FM demodulator
ang = angle(inp1);
dph = diff(ang);
dph(dph < -pi) = dph(dph < -pi) + 2*pi;
dph(dph > pi) = dph(dph > pi) – 2*pi;
fff_source = dph*Fs/2/pi;

mm = max(abs(fff_source));
sound(0.2* fff_source /mm,Fs);

% Interference
f1 = 5000;
f2 = 9000;
interf = 5000*(exp(j*2*pi*f1*t) + exp(j*2*pi*f2*t));
[b,a] = butter(2,12000/(Fs/2));
sig = filter(b,a,interf);
sig_mp = [zeros(offset,1); sig];
sig_mp = sig_mp(1:length(sig));
noise = 310*exp(j*2*pi*randn(length(sig),1));

% Combined source and interference
inp2 = inp1 + sig + sig_mp + noise;

% Signal to interference ratio
sig_2_interf = db(std(inp1)/std(sig + sig_mp));

% Signal to noise ratio
snr = db(std(inp1)/std(noise));
```

```
% FM demodulator
ang = angle(inp2);
dph = diff(ang);
dph(dph < -pi) = dph(dph < -pi) + 2*pi;
dph(dph > pi) = dph(dph > pi) – 2*pi;
fff_disturbed = dph*Fs/2/pi;

mm = max(abs(fff_disturbed));
sound(0.2* fff_disturbed/mm,Fs);
```

We now apply and compare two algorithms to attenuate the known interference.

13.3.1   Least Mean Squares Cancelling Algorithm

The LMS algorithm MATLAB code that adapts a FIR filter to the interference response is as follows, the algorithm [1] works on the real part of the signal and assumes a similar response to the imaginary part

<u>Code 13.7</u>


```
Eq_order = 32;
KMu = 0.01;
Mu = KMu/ Eq_order / var(interf);
W = zeros(Eq_order,1);  % Equalizer

for n = Eq_order:N

  D = real(interf(n:-1 : (n – Eq_order + 1)));
  est = sum(D.*W);
  e(n) = real(inp2(n)) – est;
  W = W + Mu*e(n)*D;

end
```

The KMu coefficient values for which the algorithm converges are 0.004–0.15, after convergence of W the interference is removed from the combined signal by

```
out2 = inp2 – filter(W,1,interf);
```

```
% Relative reconstruction error
reconstruction_relative_err = db(std(inp1 - out2)/std(inp1));
```

Sig_2_interf for code 13.6 is -0.2 db and code 13.7 renders reconstruction_relative_err of -34.6 db for KMu = 0.01 and when snr >= 50 db, so the algorithm attenuates more than 34 db of the interference without noise, when snr decreases then the interference attenuation decreases accordingly and for snr below ~30 db reconstruction_relative_err converges to the snr value, a result which complies with theoretical bounds.

The source, disturbed and undisturbed audio signals after demodulation are as follows (Fig. 13.5):

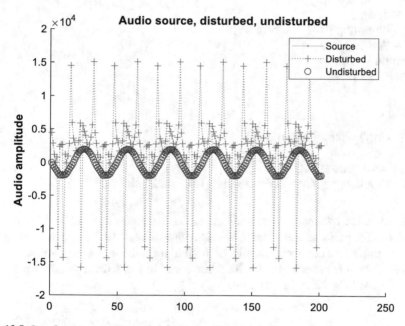

Fig. 13.5 Interference cancelling using LMS algorithm

13.3.2   *Recursive Least Squares Cancelling Algorithm*

The Recursive Least Squares (RLS) algorithm MATLAB code that adapts a FIR filter to the interference response is as follows, the algorithm [1] works on the real part of the signal and assumes a similar response to the imaginary part

Code 13.8

```
Eq_order = 32;
P = eye(Eq_order,Eq_order);
W = zeros(Eq_order,1);  % Equalizer

for n = Eq_order:N

  u = real(interf(n:-1:(n - Eq_order + 1)));
  pi_ = u'*P;
  k = 1 + sum(u'.*pi_);
  K = pi_'/k;
  e(n) = real(inp2(n)) - sum(u.*W);
  W = W + K*e(n);
  PPrime = K*pi_;
  P = P - PPrime;

end

out2 = inp2 - filter(W,1,interf);

% Relative reconstruction error
reconstruction_relative_err = db(std(inp1 - out2)/std(inp1));
```

Interference attenuation performance under the scenario of code 13.6 is very similar to the performance of the LMS algorithm at Sect. 13.3.1 above, but regarding the implementation complexity of both algorithms and the processing time which is about 3 times larger for RLS compared to LMS using MATLAB tic-toc pair, the LMS algorithm is preferable.

13.4   Conclusions

In this chapter adaptive filters were described, reviewing equalizers and interference cancellation algorithms, the details of LMS, NLMS, and RLS algorithms were described and examples were presented.

The order of the required channel estimation is not known in advance, but consecutive zeros at either side of W after convergence imply that a too high order was chosen.

Additional algorithms for adaptive filtering may be found [1] such as fast LMS, affine projection (APA), Recursive-LSL and QRD-LSL, besides being more complicated to implement than the ones we described and more processing time consumers, we did not find an advantage to use them for the tasks we described.

Reference

1. Sundar Sankaran and A.A (Louis) Beex, DSP Research Laboratory, Dept. of Electrical Engineering, Virginia Tech, Blacksburg VA 24061-0111

Chapter 14
Unequally Spaced Sampling

Abstract In this chapter we describe the Lomb–Scargle periodogram algorithm [1] [2] used for unequally spaced sampling, when certain samples of the input data are missing or unreliable, a modified algorithm is presented that shows better performance than the MATLAB internal function plomb().

Keywords Unequal · Lomb–Scargle · Periodogram · Quality factor · Plomb

14.1 Introduction

In this chapter we will describe the Lomb–Scargle periodogram algorithm [1, 2] used for unequally spaced sampling, when certain samples of the input data are missing or unreliable, or when short sampling segments should be analyzed and identified. We use a modified algorithm that shows better performance relative to the MATLAB internal function plomb().

14.2 Modified Lomb–Scargle Algorithm

Assume that we have to identify in the shortest time a tone frequency out of a large group with a difference of 2 Hz from each other, with an unknown DC component and sampled at 8192 Hz. If we try to use an FFT with a 2 Hz resolution with partially zeroed samples we get the following figure at different SNRs (Fig. 14.1).

© The Author(s), under exclusive license to Springer Nature Switzerland AG 2022
A. Dickman, *Verified Signal Processing Algorithms in MATLAB and C*,
https://doi.org/10.1007/978-3-030-93363-0_14

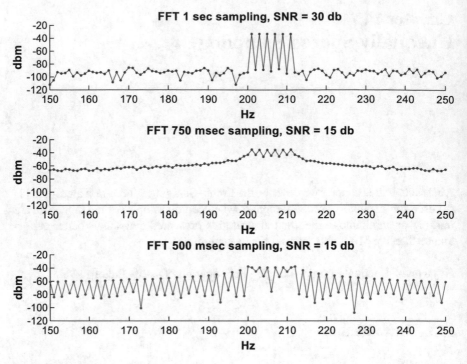

Fig. 14.1 Tone frequency estimation using FFT

As we can see a sampling interval of 500–750 mS is required in order to identify correctly the right frequency, using a window in order to improve the identification has no advantage since any window widens the bins and obstructs the identification.

As will be seen, the following modified Lomb–Scargle periodogram algorithm [1, 2] provides a solution to the above problem using a much shorter sampling time.

The algorithm estimates also the DC offset of the input signal, and uses it for improvement, in order to do it the algorithm minimizes the observation number of samples as close as possible to a whole number of full cycles of the investigated frequency, the algorithm accommodates with a sampling interval which does not include even a full cycle of the signal's frequency.

The algorithm inputs are the samples and the sampling times, these and other parameters are defined in MATLAB code as follows:

Code 14.1

```
Tm = 0.02;
Fs = 8192;
Test_Amp = 0.4;
Test_freqs = (80:2:278)';
MT = 100;
fr = repmat(Test_freqs',MT,1);
j = sqrt(-1);
N = round(Tm*Fs);
xt = [1:N]'/Fs;
DC_in = 0.6;
SNR = 15;
```

The main calculations of the algorithm may be done offline as follows, based [3] on the data in code 14.1

Code 14.2

```
% Prepare offline data
for k = 1:length(Test_freqs)

  L1(k) = floor(N*Test_freqs(k)/Fs);
  L1(k) = round(L1(k)*Fs/Test_freqs(k));
  if (L1(k) == 0)
    L1(k) = N;
  end
  Head1(k) = ceil((N – L1(k))/2);
  Tail1(k) = N – Head1(k) – L1(k);

end

cc = exp(j*2*pi*xt*Test_freqs');
sumsh = 2*sum(imag(cc).*real(cc));
sumc = sum((real(cc) – imag(cc)).*(real(cc) + imag(cc)));
wtau = 0.5*atan(sumsh./sumc);
tau = wtau./(2*pi*Test_freqs');
swtau = sin(wtau);
cwtau = cos(wtau);
ti = repmat(xt,1,length(Test_freqs)) – repmat(tau,length(xt),1);
ci = exp(j*2*pi*ti.*repmat(Test_freqs',size(ti,1),1));
```

The online part of the algorithm including the test signal preparation in MATLAB code is as follows [3]:

Code 14.3

```
est_freq = zeros(MT,length(Test_freqs));
est_freq_plomb = est_freq;
est_DC = est_freq;
Max_period = est_freq;

for i = 1:length(Test_freqs)

  f = Test_freqs(i);

  for m = 1:MT

    dp = 2*pi*m/MT;
    x1 = Test_Amp*sin([1:N]'*2*pi*f/Fs + dp);
    noise = randn(length(x1),1);
    Pn = sum(noise.^2);
    Ps = sum(x1.^2);
    scale = sqrt(Ps/Pn)*10^(-SNR/20);
    noise = scale*noise;
    x1 = x1 + noise + DC_in;

% Calculate cumsum
    cum = zeros(1,length(Test_freqs) + 1);
    for k = 2 : (length(x1) + 1)
      cum(k) = cum(k – 1) + x1(k – 1);
    end

    sumx1 = 2*sum(x1);
    sumx12 = sum(x1.^2);

    for k = 1:length(Test_freqs)

      mab(k) = (cum(N – Tail1(k) + 1) – cum(Head1(k) + 1))/L1(k);
      vary(k) = (sumx12 – sumx1*mab(k) + N*mab(k).^2)/ (N – 1);

    end

    sumcy = sum(real(ci).*repmat(x1,1,size(ci,2))) – mab.*sum(real(ci));
    sumsy = sum(imag(ci).*repmat(x1,1,size(ci,2))) – mab.*sum(imag(ci));
    py = ((sumcy.^2./sum(real(ci).^2)) + (sumsy.^2./sum(imag(ci).^2)))/1./vary;

% Frequency estimation
    [maxpy,ix] = max(py);
    est_freq(m,i) = Test_freqs(ix);
```

```
% MATLAB plomb function
   [P1,f1] = plomb(x1,xt,Test_freqs);
   imax1 = find(P1 >= max(P1));
   est_freq_plomb(m,i) = f1(imax1);
% DC estimation
   est_DC(m,i) = mab(ix);
% Max periodogram
   Max_period(m,i) = sqrt((maxpy/N));

  end
end
```

14.3   Algorithm Performance

The estimated frequencies, test frequencies, and the MATLAB plomb() estimation
for SNR = 15 db and sampling interval of 20 mS are shown in the following figure
along with the DC estimation (Fig. 14.2).

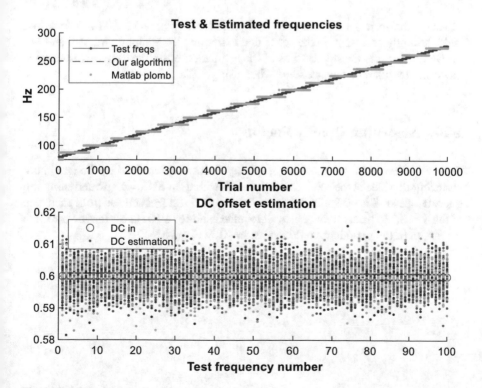

Fig. 14.2 Lomb–Scargle algorithm performance

If we compare the estimation error bias and std of our algorithm with the MATLAB plomb() function for different SNRs and sampling intervals, we get the following table (Table 14.1).

Table 14.1 Lomb–Scargle algorithm performance using our algorithm and MATLAB plomb

Algorithm	SNR [db]	Sampling interval [mS]	Error bias [Hz]	Error std [Hz]
Our Algorithm	15	10	0.54	2.33
Plomb	15	10	0.065	3.28
Our Algorithm	15	15	0.025	0.64
Plomb	15	15	0.005	1.09
Our Algorithm	15	20	0.004	0.2
Plomb	15	20	0.046	0.49
Our Algorithm	25	10	0.46	1.97
Plomb	25	10	0.075	3.06
Our Algorithm	25	15	0	0
Plomb	25	15	-0.02	0.86
Our Algorithm	25	20	0	0
Plomb	25	20	0.05	0.38

As can be seen our algorithm is preferable to the plomb() MATLAB function, and is strongly consistent in the sense that increasing the sampling interval and / or the SNR decreases the std and bias of the estimation error, the sampling interval is much shorter than the one specified when using FFTs.

14.4 Algorithm Quality Factor

The estimation reliability measure or quality factor is determined by the maximum periodogram value at the estimated frequency, which for SNR = 15 db and sampling interval of 20 mS is ~0.98, in the following figure 5 test frequencies from each side of the 80:2:278 Hz test frequencies were added and for a 10 Hz estimation error the maximum of the periodogram decreases to ~0.9 (Fig. 14.3).

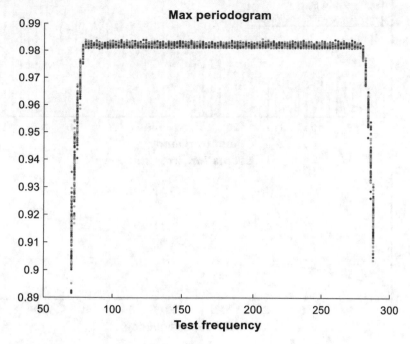

Fig. 14.3 Lomb–Scargle algorithm quality factor

14.5   Algorithm Sensitivity to Interference

The algorithm sensitivity to an interefering frequency for the example above, with the same amplitude of the tested frequency was checked, we checked the effect on the case of SNR = 15 db and sampling interval of 20 mS, for which an error std of 0.2 Hz was achieved.

The error std of the lower half estimation of Test_freqs to a varying frequency between 5 and 50 Hz was investigated as follows (Fig. 14.4):

Next the error std of the upper half estimation of Test_freqs to a varying frequency between 300 and 450 Hz was investigated as follows (Fig. 14.5):

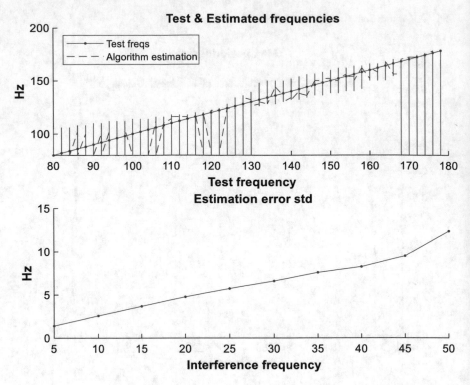

Fig. 14.4 Lomb–Scargle algorithm sensitivity to interference from below

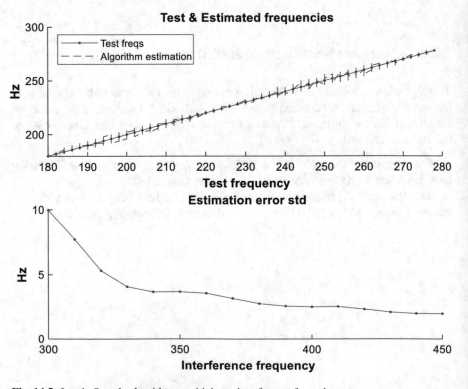

Fig. 14.5 Lomb–Scargle algorithm sensitivity to interference from above

As may be seen, an interference frequency from below has relatively a higher impact on the error std, as the number of cycles of the tested lower frequencies within the sampling interval is lower than for the higher ones.

14.6 Algorithm Implementation in C

As explained in Sect. 14.2 part of the algorithm may be calculated offline using tables so real time resources are less exploited, the following variables which depend on sampling interval and test frequencies are saved in an *.h file as elaborated in the following table, the variables refer to Code 14.1.

The ci elements are arranged as an interleaved array (real(ci(0)), imag(ci(0)), ... real(ci(N-1)), imag(ci(N-1)), all for test frequency 1, then the other arrays for the rest of the test frequencies (Table 14.2).

Table 14.2 Lomb–Scargle algorithm variables used for implementation

Variable	Size	Type
Num_test_freqs	1	int
Test_freqs	Num_test_freqs	int
N	1	Int
Ci	2*N*Num_test_freqs	float
Sum(real(ci))	Num_test_freqs	float
Sum(imag(ci))	Num_test_freqs	float
1./Sum(real(ci).^2)	Num_test_freqs	float
1./sum(imag(ci).^2)	Num_test_freqs	float
1./L1	Num_test_freqs	float
Head1	Num_test_freqs	int
Tail1	Num_test_freqs	int

A C code to implement the online code 14.3 is as follows:

Code 14.4

```c
void lomb_scargle(float *x, int n, int *pEst_freq, float *pEst_DC, float
*pEst_py)
{
int k, j;
float cum[1024], mab[1024], vary[1024];
float sumx1, sumx12, sumcy, sumsy, py, max_py;

cum[0] = (float)0.;
for (k = 1; k < n + 1; k++)
  cum[k] = cum[k - 1] + x[k - 1];

sumx1 = (float)0.;
sumx12 = (float)0.;
for (k = 0; k < n; k++)
  {
  sumx1 += x[k];
  sumx12 += (x[k] * x[k]);
  }
sumx1 *= (float)2.;

for (k = 0; k < Num_Test_freqs; k++)
  {
  mab[k] = (cum[n - Tail1[k]] - cum[Head1[k]]) * inv_L1[k];
  vary[k] = (sumx12 - sumx1*mab[k] + n*mab[k]*mab[k]) / (float)(n - 1);
  }

max_py = (float)0.;
*pEst_freq = 0;
*pEst_DC = (float)0.;
*pEst_py = (float)0.;
for (k = 0; k < Num_Test_freqs; k++)
  {
  sumcy = (float)0.;
  sumsy = (float)0.;
  for (j = 0; j < n; j++)
    {
    sumcy += ci[2*j + 2*k*n]*x[j];
    sumsy += ci[2*j + 1 + 2*k*n]*x[j];
    }

  sumcy -= (mab[k]*sum_real_ci[k]);
  sumsy -= (mab[k]*sum_imag_ci[k]);
```

```
py = (sumcy*sumcy*inv_sum_real_ci2[k] + sumsy*sumsy*
inv_sum_imag_ci2[k]) / vary[k];

if (py > max_py)
  {
  max_py = py;
  *pEst_freq = Test_freqs[k];
  *pEst_DC = mab[k];
  *pEst_py = max_py / (float)n;
  }
 }
}
```

The processing time for code 14.4 on an I5 2.9 GHz CPU is 548 nS per sample on 100 test frequencies, each doubling of the number of test frequencies doubles also the processing time per sample.

14.7 Conclusions

In this chapter we described a modified Lomb– Scargle periodogram algorithm [1] [2] used for unequally spaced sampling, and demonstrated it to identify a tone frequency with an accuracy of ±1 Hz out of 100 tones using a much lower sampling interval than required when using a conventional FFT. When comparing the proposed algorithm to the MATLAB built in function plomb(), the proposed algorithm shows better performance. More details and background information about the algorithm may be found in [4].

References

1. Lomb NR (1976) Least-squares frequency analysis of unequally spaced data. Astrophys Space Sci 39:447–462
2. Scargle JD (1982) Studies in astronomical time series analysis. II. Statistical aspects of spectral analysis of unevenly spaced data. Astropys J 302:757–763
3. Numerical Recipes in C (1992) William H. Press, Saul A. Teukolsky, William T. Vetterling, Brian P. Flannery: 575–581.
4. Understanding the Lomb-Scargle periodogram (2018) Jacob T. VanderPlas, The astrophysical journal supplement series, 236:16, 2018.

Chapter 15
Encountered Problems in Signal Processing & Solutions

Abstract In this chapter we describe the following encountered problems in signal processing and how to solve them: Input signal spikes, fast input transitions, improving processing time, center frequency estimation, power level estimation, noise floor estimation, bandwidth estimation, automatic level control, frequency response estimation, group delay estimation.

Keywords Spikes · Transitions · WOLA · dbm · Center frequency · Multiple · Power · Noise floor · Bandwidth · Estimation · ALC · Frequency response · Unwrap · Group delay · Estimation

15.1   Introduction

In this chapter we will describe encountered signal processing problems and how to solve them.

15.2   When Input Signal Includes Spikes

Sampling in a noisy or hostile environment may suffer from sudden interruptions or spikes occurring randomly, which raise the noise floor of a performed FFT and therefore decrease the dynamic range.

A MATLAB model of such a signal and its FFT following WOLA 3 window is

© The Author(s), under exclusive license to Springer Nature Switzerland AG 2022   173
A. Dickman, *Verified Signal Processing Algorithms in MATLAB and C*,
https://doi.org/10.1007/978-3-030-93363-0_15

Code 15.1

```
Fs = 12.8e6;
Nf = 2048;
N3 = 3*Nf;
t = (0:N3-1)/Fs;
Fin = 2e6;
sig = 100*cos(2*pi*Fin*t);

win_cheb = chebwin(N3,106)';
win_cheb = win_cheb / max(win_cheb);

sigt1 = win_cheb.*sig;
sigt1 = sigt1(1:Nf) + sigt1(Nf + (1:Nf)) + sigt1(2*Nf + (1:Nf));
Fout1 = fft(sigt1,Nf)/Nf;
Fout1 = Fout1(1:Nf/2);

sig1 = sig;

% spike
sig1(2014) = 50*sig1(2014);

sigt2 = win_cheb.*sig1;
sigt2 = sigt2(1:Nf) + sigt2(Nf + (1:Nf)) + sigt2(2*Nf + (1:Nf));
Fout2 = fft(sigt2,Nf)/Nf;
Fout2 = Fout2(1:Nf/2);
```

As seen at the following figure, a single spike at one sample out of 6172, magnifying it by 50, raises the noise floor by about 70 db, while strong signals stay unchanged, if that remains then only signals which are larger than 30 db below the strong signal will be detected (Fig. 15.1).

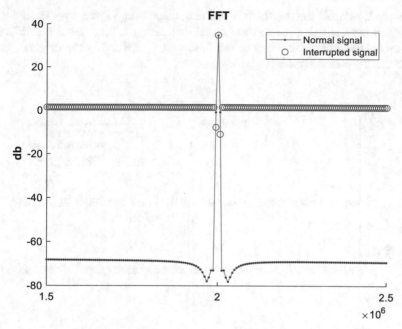

Fig. 15.1 Spikes effect before solution

A solution to the problem is to clip each sample to some level above a measure of the mean amplitude at a signal segment prior to the current segment, in the following MATLAB code, clipping to 4 times that measure improves the situation by 30 db

Code 15.2

```
L = 128;
G = 4;
Th = G*mean(abs(sig(1:L)));

for k = 1:length(sig1)

 if (abs(sig1(k)) > Th)
  sig1(k) = sig1(k)*Th/abs(sig1(k));
 end

 if (mod(k,L) == 0)
  Tcurr = G*mean(abs(sig1((k - L + 1):k)));
  Th = 0.75*Th + 0.25*Tcurr;
 end

end
```

Every L samples the threshold is updated, decreasing G improves the dynamic range but has the risk to distort the signal, increasing L may lower the ability to track dynamic amplitude changes, the following figure shows the original, interrupted, and repaired signals (Fig. 15.2).

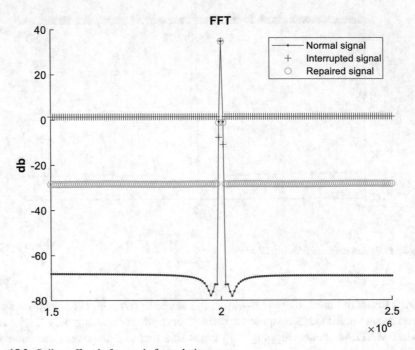

Fig. 15.2 Spikes effect before and after solution

A C implementation of code 15.2 is

Code 15.3

```c
void clip(int *x, int n, int lim, int Log)
{
int k, m, sum1, temp, Th, Th_curr, mask;

mask = (2 << Log) - 1;
sum1 = 0;
for (k = 0; k < (mask + 1); k++)
  {
  if (x[k] < 0)
    sum1 -= x[k];
  else
    sum1 += x[k];
  }
sum1 >>= Log;

Th = lim*sum1;
for (k = 0; k < n; k++)
  {
  temp = x[k];
  if (temp < 0)
    temp = -temp;

  if (temp > Th)
    x[k] *= Th/temp;

  if ((k & mask) == 0)
    {
    sum1 = 0;
    for (m = 0; m < (mask + 1); m++)
      {
      if (x[k + m] < 0)
        sum1 -= x[k + m];
      else
        sum1 += x[k + m];
      }
    sum1 >>= Log;

    Th_curr = lim*sum1;
    Th = ((3*Th) >> 2) + (Th_curr >> 2);
    }
  }
}
```

Log at the C code is log2(L) at the MATLAB code 15.2, the processing time for code 15.3 on an I5 2.9 GHz CPU is 7 nS per sample.

15.3   When Input Signal Includes Fast Transitions

Fast transitions at the input signal cause spectral expansion and raise the neighboring bins to the active bins, masking the ability to detect low power activity at those bins, a solution is to use a WOLA window that smoothes the transitions, higher WOLA order decreases spectral expansion, in the following figure we show the phenomena for different conditions (Fig. 15.3)

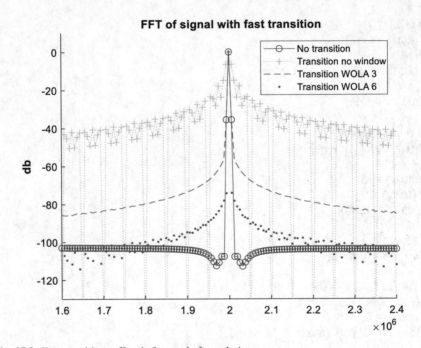

Fig. 15.3 Fast transitions effect before and after solution

WOLA 3 description in MATLAB was discussed in Sect. 3.2.2, a MATLAB description of WOLA 6 window is as follows:

Code 15.4

```
Fs = 12.8e6;
Nf = 2048;
N6 = 6*Nf;
t6 = (0:N6-1)/Fs;
sig6 = 100*cos(2*pi*Fin*t6);
sig6(1:N3/8) = 0;      % Transition
win_cheb6 = chebwin(N6,106)';
sigt6 = win_cheb6.*sig6;

sigt6 = sigt6(1:Nf) + sigt6(Nf + (1:Nf)) + sigt6(2*Nf + (1:Nf)) +...
   sigt6(3*Nf + (1:Nf)) + sigt6(4*Nf + (1:Nf)) + sigt6(5*Nf + (1:Nf));
Fout6 = fft(sigt6,Nf)/Nf;
Fout6 = Fout6(1:Nf/2);
Amp_Fout6 = abs(Fout6).^2;
Amp_Fout6 = Amp_Fout6/max(Amp_Fout6);
```

Another solution would be to slow the rise time of the signal but this is not always applicable.

15.4   Algorithms Processing Time Improvement

When implementing an algorithm in software, mathematical operations such as log(), exp(), sqrt(), and others may be required, these operations consume relatively high processing time which may be reduced by using direct access to tables, nowadays memory is not a critical issue for SBCs or even DSPs so that this attitude may be preferable on using expensive FPGAs to implement the algorithm.

Consider to calculate the total power of n signals whose power is given in dbm with a resolution of 0.1 dbm, the straight forward way would be to make $\log 2(n)$ operations of conversion to decimal units, add all and convert to dbm, a MATLAB description of this way for 2 signals is

Code 15.5

```
p1 = 90;
p2 = 91.4;

p1n = 0.001*10^(p1/10);
p2n = 0.001*10^(p2/10);

sump_real = 10*log10((p1n + p2n)/0.001);
```

Implementing code 15.5 in C for just the 10^ operations which obligates to include the math.h file is

Code 15.6

```
double add_dbm_dir(double x1, double x2)
{

double b1,b2;

b1 = (double)0.001 * exp((double)0.23026*x1);
b2 = (double)0.001 * exp((double)0.23026*x2);

return (b1 + b2);

}
```

This code uses processing time of 145 nS on an I5 2.9 GHz CPU.

The above operation may be approximated using the following algorithm in MATLAB code

<u>Code 15.7</u>

```
t = 0:0.1:99;
p1 = 40;
p2 = p1 - t;

p1n = 0.001*10^(p1/10);
p2n = 0.001*10.^(p2/10);
sump_real = 10*log10((p1n + p2n)/0.001);

sump_app = zeros(size(sump_real));
for i = 1 + round(10*t)
  sump_app(i) = max(p1,p2(i));
end

% Prepare correction table
del = sump_real - sump_app;
xx = find(abs(del) >= 0.08);
corr_tab = del(xx);
corr_tab = corr_tab(1:4:end);

% Algorithm
sump_dbm = zeros(size(sump_real));
for i = 1 + round(10*t)
  del(i) = max(p1,p2(i)) - min(p1,p2(i));

  deli = 1 + round(10*del(i));

  if (deli >= 170)
    sump_dbm(i) = max(p1,p2(i));
  else
    mod1 = mod(deli,4);
    if (mod1 < 1)
      deli = deli/4 + 1;
    else
      deli = round((mod1 + deli)/4);
    end
    sump_dbm(i) = max(p1,p2(i)) + corr_tab(deli);
  end

end
```

corr_tab is a 44 floating point elements table, the sum of 2 dbm values depends on the maximum and difference between these values, we chose 40 dbm to be the highest power for this example.

The accuracy of the algorithm is demonstrated on the following figure, the error does not exceed 0.1 db (Fig. 15.4).

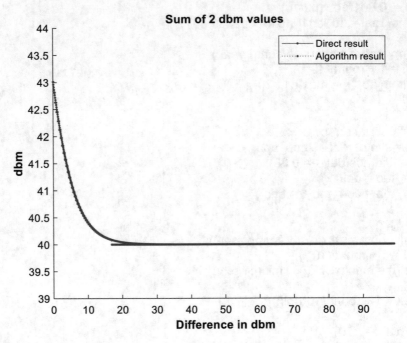

Fig. 15.4 Real and algorithm results for summing 2 dbm values

The C implementation of code 15.7 is

Code 15.8

```c
float add_dbm_alg(float x1, float x2, const float *cor)
{
int deli, mod1;
float del, max1;

if (x1 >= x2)
  {
  del = x1 - x2;
  max1 = x1;
  }
else
  {
  del = x2 - x1;
  max1 = x2;
  }
deli = (int)((float)10.*del + (float)0.5);

if (deli >= 170)
  return (max1);
else
  {
  mod1 = (deli & 3);
  if (mod1 < 1)
   deli = (deli >> 2) + 1;
  else
   {
   deli = (mod1 + deli) >> 2;
   if ((deli % 4) >= 2)
     deli += 1;
   }

  return (max1 + corr_tab[deli]);
  }
}
```

This code uses processing time of 32 nS on an I5 2.9 GHz CPU, which is about 5 times lower than the direct method, both need log2(n) such operations to calculate the addition of n signals power given in dbm.

15.5   Mean Center Frequency Estimation of Signals

Center frequency estimation is a property of the signal often required, whether a single or multiple signals are collected by the receiver, in the following we differ between the 2 cases

15.5.1   Mean Center Frequency Estimation for Multiple Signals

When multiple signals are present at the received frequency range, FFT is a tool to analyze the center frequency of each signal, which is the local maximum at the respective bin, since the measurement resolution is Fs/N_{fft} and N_{fft} is not unlimited, then this method has a limited resolution, a way to improve that is to use an NCO and LPF explained in Sects. 3.1.2 and 3.1.4 in order to lower the sampling frequency and therefore improve the measurement resolution, but that requires additional computations.

In the following MATLAB code, 3 FM signals with the same modulation but different initial phases are sampled and use a WOLA 3 window before FFT

Code 15.9

```
Fs = 12.8e6;
Nf = 2048;
res = Fs/Nf;
N3 = 3*Nf;
j = sqrt(-1);
win_cheb = chebwin(N3,106)';

t = (0:N3-1)/Fs;

F1 = 2e6;
F2 = F1 + 10*res;
F3 = F1 - 10*res;

off1 = 1000;
Faudio = 200;
Fdev = 2000;

% FM modulation
tempMod1 = Fdev*cumsum(sin(2*pi*Faudio*t))/Fs;
tempMod2 = Fdev*cumsum(sin(2*pi*Faudio*t + pi/3))/Fs;
tempMod3 = Fdev*cumsum(sin(2*pi*Faudio*t + pi))/Fs;

sig1 = 100*exp(j*2*pi*((F1 + off1)*t + tempMod1));
sig2 = 100*exp(j*2*pi*((F2 + off1)*t + tempMod2));
sig3 = 100*exp(j*2*pi*((F3 + off1)*t + tempMod3));
sig = sig1 + sig2 + sig3;

% Make WOLA 3
sigt1 = win_cheb.*sig;
sigt1 = sigt1(1:Nf) + sigt1(Nf + (1:Nf)) + sigt1(2*Nf + (1:Nf));
Fout = fftshift(fft(sigt1,Nf))/Nf;
Amp_Fout = abs(Fout).^2;
Amp_Fout = Amp_Fout / max(Amp_Fout);
```

If the three signals would have been non-modulated, which is usually not the case, then using the power of the neighbors to the local maxima and a suitable mapping of the WOLA 3 window response versus the frequency offset off1, the frequency estimation result could be significantly improved. But as for a modulated signal the local maxima locations vary at different instances, then this direction cannot be used.

For measuring the instantaneous frequency offset, the power of the neighbors could be beneficial.

In the following figure, we show the FFT of the 3 signals above along with 3 non-modulated signals at the same frequencies, starting at different initial phases, as may be seen, although the non-modulated signals power at the local maxima is different, the difference between the power of neighbors relative to local maxima is the same and could be used for center frequency estimation (Fig. 15.5).

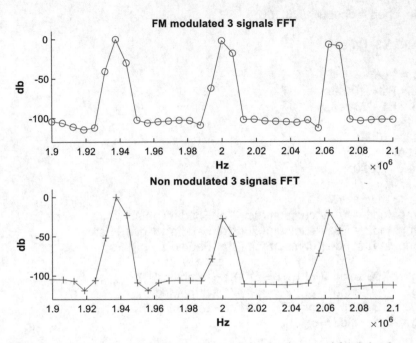

Fig. 15.5 Center frequency estimation for multiple modulated and non-modulated signals

15.5.2   *Mean Center Frequency Estimation for a Single Channel*

When it is known that just one signal is present within the received frequency range, a different algorithm may be used to estimate the mean of center frequency, this algorithm may also be applied to a signal that was isolated from a wide band stream using an NCO and an appropriate filter.

An algorithm to estimate the center frequency of an FM modulated signal is as follows:

Code 15.10

```
Fs = 2e6;
j = sqrt(-1);

F1 = 1e5;
Off1 = 2000;
Faudio = 2000;
Fdev = 10000;
N = round(10*Fs/Faudio);
t = (0:N-1)/Fs;


% Signal
tempMod = Fdev*cumsum(sin(2*pi*Faudio*t))/Fs;
sig1 = round(10000*exp(j*2*pi*((F1 + off1)*t + tempMod)));

% FM demodulator and mean center frequency estimator
ang = angle(sig1);
dph = diff(ang);
dph(dph < -pi) = dph(dph < -pi) + 2*pi;
dph(dph > pi) = dph(dph > pi) - 2*pi;
fff = dph*Fs/2/pi;

Freq_est = mean(fff);
```

The mean center frequency is composed of the RF frequency F1 and the signal offset frequency off1, reasonable accuracy is achieved when the sampling interval includes at least 10 cycles of Faudio, this algorithm may also be used as an FM demodulator by hearing fff through a sound device.

In order to implement in C code 15.10, we use a 2049 type short elements $\tan^{-1}$(val) table in the range 0 to pi/4 for the angle function, where val is between 0 and 1, the MATLAB code to implement is as follows:

Code 15.11

```
% pi/4 ←→ 32000
range = 2048;
for k = 1:(range + 1)
  at_tab(k) = round(128000/pi*atan((k - 1)/range));
end
```

Implementing code 15.10 in C is as follows

Code 15.12

```
#define Fs_div_2pi (float)2.e6 / (float)256000.

int fm_dem(int *x, int n, int Fs, const unsigned short *at, int *y)
{
int k, abs1, abs2, Last_phase, phase, temp, acc_phase;

acc_phase = 0;
Last_phase = 0;
for (k = 0; k < n; k++)
 {
 abs1 = x[2*k + 1];
 if (abs1 < 0)
   abs1 = -abs1;
 abs2 = x[2*k];
 if (abs2 < 0)
   abs2 = -abs2;

 if (abs1 <= abs2)
   phase = at_tab[(unsigned short)((float)0.5 + (float)2048.*(float)abs1/(float)abs2)];
 else
   phase = 64000 - at_tab[(unsigned short)((float)0.5 + (float)2048.*(float)abs2/(float)abs1)];

% Quadrants 2 + 3
 if (x[2*k] < 0)
   phase = 128000 - phase;

% Quadrant 4
 if (x[2*k + 1] < 0)
   phase = -phase;

 temp = phase;
 phase -= Last_phase;
 Last_phase = temp;

% Unwrap
 if (phase < -128000)
   phase += 256000;
 if (phase > 128000)
   phase -= 256000;

 y[k] = (int)((float)0.5 + (float)phase * Fs_div_2pi);
 acc_phase += phase;

 }
return((int)((float)0.5 + (float)acc_phase * Fs_div_2pi / (float)n));
}
```

Code 15.12 uses a processing time of 31 nS per sample on an I5 2.9 GHz CPU.

15.6   Power Level Estimation

Power level estimation is often required for applying automatic level control (ALC) or to distinguish between signals, we also differ between single and multiple signals collected by a receiver, the power level is usually measured by dbm, defined by

$P_{dbm} = 10*log10(P / 0.001)$

where P is the power in Watts

Power computation for a single signal in time domain using MATLAB is as follows:

Code 15.13

```
Fs = 5.12e6;
N = 8192;
t = (0:N-1)/Fs;
res = Fs/N;
F1 = 1000*res;
sig = 32767*sin(2*pi*F1*t) ;

Kp = 1/0.5/(32767^2);
P = Kp*sum(abs(sig).^2)/N;
Pdbm = 10*log10(P);
```

The constant Kp is determined by normalization to a known dbm measurement for a certain input level, code 15.13 assumes that a full scale 16 bits signed format is expected to be measured as 0 dbm at the a2d output.

The signal power may be estimated through the frequency domain as well using the following MATLAB code:

Code 15.14

```
ind = 1001;
Fout = fft(sig,N)/N;
Fout = abs(Fout).^2;
Pfft = Kp*sum(Fout);
Pfft_dbm = 10*log10(Pfft);

Pfft_bin = Kp*Fout(ind);
Pfft_bin_dbm = 10*log10(Pfft_bin);
```

The total frequency domain power Pfft is equal to the total time domain power P, in accordance with Parseval's theorem, the power Pfft_bin within the relevant bin is

½ of Pfft since sig is a real signal so its fft is symmetric around 0 Hz and the signal power is divided between the positive and negative frequencies.

As seen from the following figure (Fig. 15.6), a non-modulated, FM modulated, and AM modulated signals are presented, with no shaping window used, for the

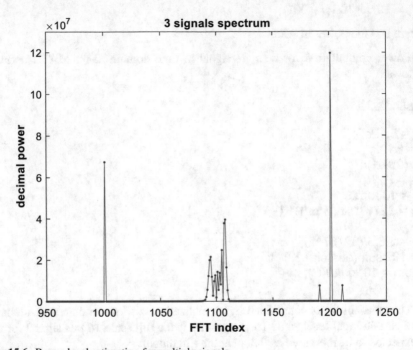

Fig. 15.6 Power level estimation for multiple signals

modulated signals or if a window would have been used, then summing the additional bins power around the local maxima is required in order to compute the signals power as follows:

Code 15.15

```
ind1 = 1001;
ind2 = 1101;
ind3 = 1201;

Pfft1 = Kp*Fout(ind1)
Pfft2 = Kp*sum(Fout((ind2 - 10):(ind2 + 10)))
Pfft3 = Kp*sum(Fout((ind3 - 10):(ind3 + 10)))
```

15.7   Noise Floor Estimation

Noise floor estimation for the received range of frequencies is required in order to estimate SNR for the different signals in the frequency range and for estimating correctly the bandwidth of a signal with low to moderate SNR. As explained in Sect. 1.2 the SNR that ensures detection needs to be at least 13 db, therefore it is important to know how close it is to that limit where its detection is no longer guarantied.

If we investigate a typical spectrum of a receiver around the IF frequency after WOLA 3 and a 8192 points instantaneous FFT, we observe that the receiver bandwidth comprises 5120 bins and 40 active signals above the noise floor as follows (Fig. 15.7):

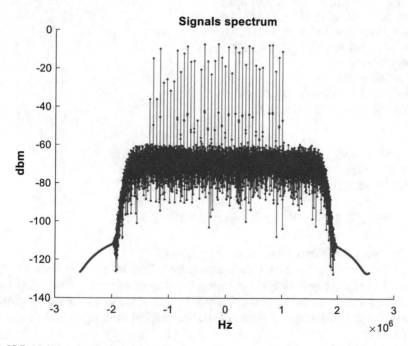

Fig. 15.7 Noisy multi signal spectrum

As most of those 5120 bins contain noise, an obvious estimate to the noise floor is the most frequent power level that appears among those 5120 bins, as estimation error of 1 db or less is sufficient for noise floor estimation we use the bins power in dbm and a histogram for estimation, the algorithm in MATLAB code is as follows:

Code 15.16

```
Fs = 5.12e6;
N = 8192;
t = (0:N-1)/Fs;
res = Fs/N;

win_cheb = chebwin(3*N,106)';
sigt1 = win_cheb.*sig;
sigt1 = sigt1(1:N) + sigt1(N + (1:N)) + sigt1(2*N + (1:N));
Fout = fftshift(fft(sigt1,N))/N;
Fout = Kp*abs(Fout).^2;

low_ind = round((Fs/2 - 1.6e6)/res);
high_ind = round((Fs/2 + 1.6e6)/res) - 1;
slice = Fout(low_ind:high_ind);
slice_db = round(10*log10(slice));
slice_db_sav = 0;
if (min(slice_db) < 0)
  slice_db_sav = min(slice_db);
  slice_db = slice_db - slice_db_sav;
end

ax = 0:max(slice_db);
his = hist(slice_db,ax);
[y,Ind] = max(his);

est_noise_db = Ind + slice_db_sav - 1;
```

The histogram of bins power in db is as follows:

If a better resolution than 1 dbm is required, then slice_db in code 15.16 is rounded to fractions of db and the histogram shape might be not as sharp as in Fig. 15.8, for this case a finer estimation may be achieved using a parabolic approximation, consider 3 histogram values of the peak and ±1 unit, then if we denote the 3 values by

$(x1,y1),(x2,y2),(x3,y3)$ where $y2 > \max\{y1,y3\}$, $x3 - x2 = x2 - x1 =$ resolution, then the peak location offset with respect to x2 is

$$\text{Offset} = \frac{(y1 - y3) * \text{resolution} / 2}{(y1 + y3 - 2 * y2)}$$

The noise floor plotted on the spectrum figure is as follows (Fig. 15.9):
Implementing code 15.16 in C is as follows:

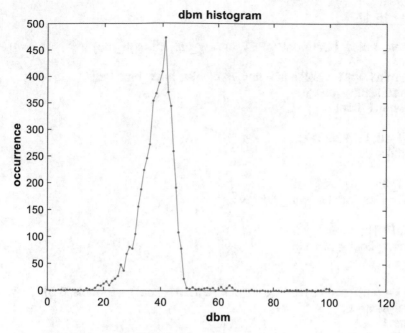

Fig. 15.8 Dbm histogram of a noisy multi signal spectrum

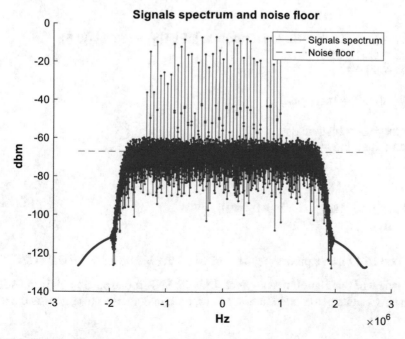

Fig. 15.9 Noisy multi signal spectrum and noise floor estimation

Code 15.17

```
int calc_noise_floor(float *pFFT, int low_ind, int high_ind)
{
int i, min1, val1, val2, min1_sav, max_hist, max_hist_ind;
int hist[128];
float min1_float;

for (i = 0; i < 128; i++)
  hist[i] = 0;

min1_float = (float)1.e9;
for (i = low_ind; i < high_ind; i++)
  {
  if (pFFT[i] < min1_float)
    min1_float = pFFT[i];
  }

min1 = (int)((double)10.*log10((double)min1_float) + (double)0.5);
min1_sav = 0;
if (min1 < 0)
  min1_sav = min1;
max_hist = 0;
for (i = low_ind; i < high_ind; i++)
  {

  val1 = (int)((double)10.*log10((double)pFFT[i]) + (double)0.5);
  val2 = val1 - min1_sav;
  hist[val2] += 1;

  if (hist[val2] > max_hist)
    {
    max_hist = hist[val2];
    max_hist_ind = val2;
    }
  }

  return (max_hist_ind + min1_sav);
}
```

Code 15.17 uses a processing time of 44 nS per bin on an I5 2.9 GHz CPU.

In view of this algorithm and Sect. 15.6 the SNR for a signal in db is the difference between its power and the noise floor, both in dbm, the noise floor is the total noise in the passband zone divided by the number of passband bins, each doubling of the FFT size reduces the noise floor by 3 db, as the total noise is divided to twice number of bins.

15.8   Bandwidth Estimation

Bandwidth estimation may be required in order to verify if transmissions use the spectrum legally, to distinguish between signals or to identify them.

If we investigate the following typical spectrum of a receiver around the IF frequency after WOLA 3 and a 8192 points instantaneous FFT, we observe 3 wide band separated signals and wish to estimate their bandwidth (Fig. 15.10).

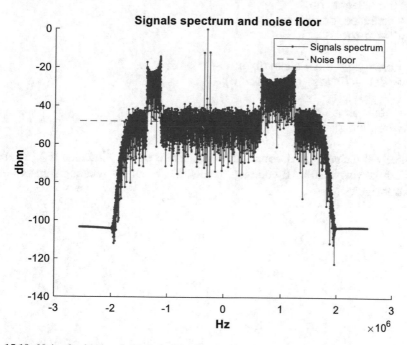

Fig. 15.10 Noisy 3 wide band signals spectrum and noise floor

Since the spectrum contains noise besides signals, it should be subtracted from the spectrum in order to measure as accurately as possible the signal itself, especially when the SNR is low to moderate, to do that we use Sect. 15.7 and part of

Code 15.17 to measure the noise floor, the computations are done on decimal values and not dbm, as in the following MATLAB code:

Code 15.18

```
slice = Fout(low_ind:high_ind);
slice_db = round(10*log10(slice));

slice_db_sav = 0;
if (min(slice_db) < 0)
  slice_db_sav = min(slice_db);
  slice_db = slice_db - slice_db_sav;
end

ax = 0:max(slice_db);
his = hist(slice_db,ax);
[y,Ind] = max(his);

est_noise_db = Ind + slice_db_sav – 1;
noise_dec = 10^(est_noise_db/10);

slice_net = slice - noise_dec;
slice_net(find(slice_net < 0)) = 0;
```

slice_net includes the decimal values of the bins power within the IF filter zone, in the following MATLAB code the bandwidths of all the separate signals in the spectrum are estimated.

Code 15.19

```
Flag = 0;
del = 80;
st = 0;
en = 0;
Flag_end = 0;
count = 1;
bw = zeros(10,1);
Th = 20;
for k = 1:(length(slice_net) - del)

  mm1 = mean(slice_net(k:(k + del - 1)));
  if (mm1 > Th*noise_dec) & (Flag == 0)
    st = k;
    Flag = 1;
    Last = mm1;
  end

  if (mm1 < Th*noise_dec) & Flag & ~Flag_end
    en = k;
    Flag = 0;
    Flag_end = 1;
  end

  if (st & en & Flag_end)
    bw(count) = (en - st - del - 2)*Fs/N;
    Flag_end = 0;
    bw_diff = en - st
    st = 0;
    en = 0;
    count = count + 1;
  end

end

bw_out =  bw(1:count);
```

del in bins should be at least the bandwidth of the widest non-populated inter-
nally signal, as for the middle signal above which is AM modulated and has a carrier
and 2 side lobes, the non-populated zone between the carrier and a side lobe has 78
bins, the threshold Th complies to 13 db.

The algorithm results may be inaccurate by 5–10% which are acceptable errors for such a measurement, they may be wrong when the signal bandwidth is less than 10% of the total bandwidth or more than 80% of it, or when the SNR is too low, as code 15.19 is straightforward the equivalent C code will not be presented.

An alternative algorithm to bandwidth estimation is to measure the bandwidth at which 97% of the total power is contained, assume the following typical spectrum (Fig. 15.11):

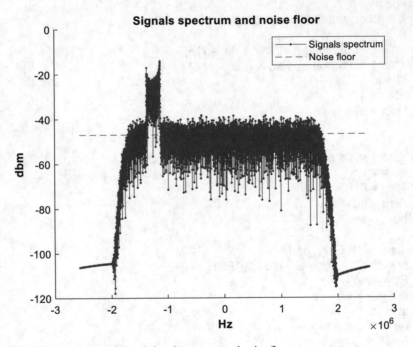

Fig. 15.11 Noisy single wide band signal spectrum and noise floor

Then an algorithm to measure the signal bandwidth is

<u>Code 15.20</u>

```
slice = Fout(low_ind:high_ind);
slice_db = round(10*log10(slice));

slice_db_sav = 0;
if (min(slice_db) < 0)
  slice_db_sav = min(slice_db)
  slice_db = slice_db - slice_db_sav;
end

ax = 0:max(slice_db);
his = hist(slice_db,ax);

est_noise_db = Ind + slice_db_sav - 1;
noise_dec = (10^(est_noise_db/10))/1;

slice_net = slice - noise_dec;
slice_net(find(slice_net < 0)) = 0;

sum_FFT = sum(slice_net);
InvEner = 1/sum_FFT;
Frac = 1 - 0.97;

k2 = 1;
k1 = length(slice_net);
Val = 0;
while (1)

  if ((Val * InvEner) >= Frac)
    break;
  end

  if (slice_net(k1) > slice_net(k2))
    Val = Val + slice_net(k2);
    k2 = k2 + 1;
  else
    Val = Val + slice_net(k1);
    k1 = k1 - 1;
  end

  if (k2 >= k1)
    break;
  end

end
```

Bandwidth = (k1 - k2 + 1)*Fs/N;

The comments about the measurement accuracy that were mentioned for the former method apply to this one as well.

As code 15.20 is also straightforward then the equivalent C code will not be presented.

15.9   Automatic Level Control

The spectrum of collected signals by a receiver has usually a variable average power over time, with more variations when the bandwidth grows as more signals may rise or disappear. Since it is desired to maximize the dynamic range of the receiver, namely to detect low and strong signals simultaneously, then the receiver's digitally controlled attenuator (DCA) is used dynamically to keep the average signal level at the a2d input to 10-12 db under the full scale of the a2d.

The reason for this level is that the input signals may have sudden peaks that may still saturate the a2d momentary and distort or compress its measurements.

The suggested way to control the average power is to use periodic measurements of the current power and update the DCA accordingly.

The common time T_s used for power measurements is such that $BW*T_s \geq 50$, the common periodic time between updates T_u is such that 40–150 processes have passed, where a process is an action such as WOLA followed by an FFT, small T_u tracks better the power envelope but updates of the DCA cause disturbances to the signal so T_u cannot be too small.

A MATLAB code that performs the proposed algorithm is as follows, typical values for Katt are 0.5 to 1

Code 15.21

```
Fs = 3.2e6;
N = Fs/2;
Nf = Fs/25000;
j = sqrt(-1);
Ts = 20e-6;
Nalc = round(Ts*Fs);
Tu = 5e-3;
Nu = round(Tu*Fs);

c1 = 0;
c2 = 0;
Ref = -12;

Att = 0;
Katt = 0.5;
Kp = 1/(0.5*32767^2);

while (c1 < (length(sig) - 3*Nf))

% ALC section
  if (c1 >= c2)

    sig2 = sig([1:Nalc] + c1);
    sig2 = real(sig2).^2 + imag(sig2).^2;

    Power_alc = 10*log10(Kp*sum(sig2)/Nalc) - Att;
    Att = Katt*(Power_alc - Ref) + Att;

    Att = round(max(Att,0));
    c2 = c1 + Nu;

  end

  sig1 = sig([1:3*Nf] + c1);
  sig1 = real(sig1).^2 + imag(sig1).^2;

  Power_in = 10*log10(Kp*sum(sig1)/(3*Nf));
  Power_out = Power_in - Att;

  Alg_wola3_fft();

  c1 = c1 + 3*Nf;

end
```

The algorithm output power for a typical input and a destination of -12 dbm is shown as follows (Fig. 15.12):

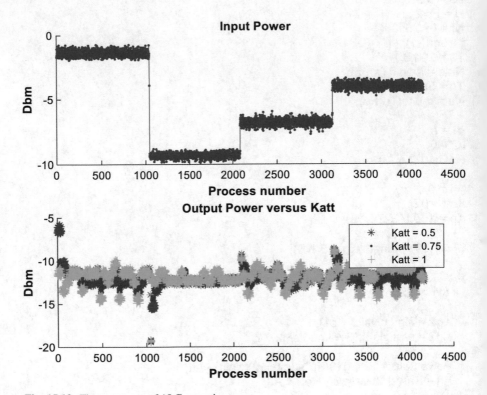

Fig. 15.12 Time response of ALC operation

A C implementation of the ALC section in code 15.21 is as follows, the input samples are in int format and interleaved

```
#include <math.h>
```

Code 15.22

```
int make_alc(int *x, int Att, int n, float Katt, int Ref)
{
int k, Att_int;
double dbm;
float sum1, Att_float;

sum1 = (float)0.;
for (k = 0; k < n; k++)
  sum1 += ((float)x[2*k]*(float)x[2*k] + (float)x[2*k+1]*(float)x[2*k+1]);

dbm = (double)10.*log10(Kp * (double)sum1 / (double)n);
Att_float = Katt*(float)(dbm - Ref);
Att_int = (int)(Att_float + (float)0.5) + Att;

if (Att_int < 0)
  Att_int = 0;

return (Att_int);

}
```

Code 15.22 uses a processing time of 312 nS on an I5 2.9 GHz CPU for each 64 samples in an ALC action.

If a secondary algorithm like demodulation is performed on the samples that undergo ALC, then a process that can improve the demodulator performance is to divide the signal to segments and multiply each complex sample by sqrt(Pref / Psig) where Pref is the reference power and Psig the power computed at the previous segment.

15.10   Frequency Response Estimation

An algorithm to estimate the amplitude and phase response of a filter or amplifier versus frequency is presented, computed from the responses of the device to a series of frequencies.

A MATLAB code to implement it is

Code 15.23

```
Fs = 16e6;
Fif = 0;
freq_vec = 1000:1000:Fs/2;
N = 4000;
t = (0:N-1)/Fs;
j = sqrt(-1);

% Tested device
[B,A] = ellip(10,0.5,100,3e6/(Fs/2));
B = B * sum(A) / sum(B);

F1 = (0:7999)*Fs/16000;
h1 = freqz(B,A,F1,Fs);

for k = 1:length(freq_vec)

  freq = Fif + freq_vec(k);
  sig1 = 100*exp(j*(2*pi*freq*t + 2*pi*rand(1)));

  sig2 = sig1.*exp(-j*2*pi*Fif*t);
  sig2 = filter(B,A,sig2);
  sig2 = sig2(1000:end);  % Ignore transient

  ref = sig1(1000:end);    % Ignore transient

% Amplitude estimation
  resp(k) = sum(abs(ref.*conj(sig2)))/sum(abs(ref).^2);

% Phase estimation
  ph(k) = angle(sum(sig2.*conj(ref)));

end

ph = unwrap(ph);
ph_freqz = unwrap(angle(h1));
```

If an IF filter at the receiver's output is tested, then Fif is set accordingly, N is chosen so that the time-bandwidth multiplication is above 1000.

Amplitude and phase responses to the example in code 15.23 are as follows (Fig. 15.13):

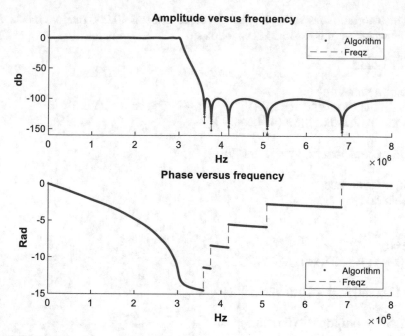

Fig. 15.13 Frequency response estimation results

Implementation of algorithm elements such as complex exponent and NCO were described in Sect. 3.1.2 code 3.1.3, and X.*conj(Y) in Sect. 12.2.1 code 12.5, a simplified MATLAB implementation to the unwrap() MATLAB function is as follows:

Code 15.24

```
% Test input
inp = -50*randn(1,1000);

% MATLAB function result
y1 = unwrap(inp);

% Simplified algorithm
y2(1) = inp(1);
acc_dp_corr = 0;
for k = 2:length(inp)
  dp1 = inp(k) - inp(k - 1);
  dp_corr1 = dp1/(2*pi);
  dp_corr1 = floor(dp_corr1 + 0.5);

  acc_dp_corr = acc_dp_corr + dp_corr1;
  y2(k) = inp(k) - 2*pi*acc_dp_corr;
end
```

The algorithm was verified by using 10,000 trials of 10,000 random inputs, a C implementation of code 15.24 is as follows:

Code 15.25

```c
#include <math.h>
#define pi2 (float)(4.*acos(0.))
#define Inv2pi (float)(1./ (4.*acos(0.)))

void unwrap(float *x, int n, float *y)
{
int k, dp_cor, acc;
double dp1;

y[0] = x[0];
acc = 0;
for (k = 1; k < n; k++)
  {
  dp1 = (x[k] - x[k - 1])*Inv2pi;

  dp_cor = (int)(dp1 + (float)0.5);

  if (((float)dp_cor - dp1) >= (float)0.5)
    dp_cor -= 1;

  acc += dp_cor;

  y[k] = x[k] - pi2*(float)acc;
  }
}
```

Unwrap() computations on a random array looks as (Fig. 15.14)

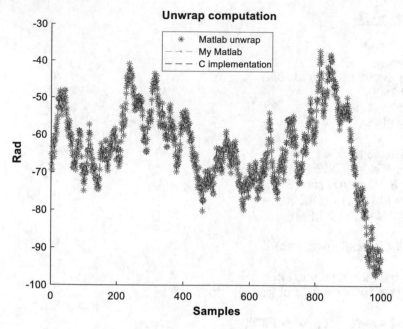

Fig. 15.14 Unwrap computation results for a random input

Code 15.25 uses a processing time of 110 nS per sample on an I5 2.9 GHz CPU.

15.11   Group Delay Estimation

An algorithm to estimate the group delay response of a filter or amplifier versus frequency is presented, computed from the responses of the device to a series of frequencies.

A MATLAB code to implement it is

Code 15.26

```
Fs = 16e6;
Fif = 4e6;
freq_vec = 1000:1000:Fs/4;
N = 3500;
t = (0:N-1)/Fs;
j = sqrt(-1);

% Tested device
[B,A] = ellip(8,0.25,80,3.7e6/(Fs/2));
B = B * sum(A) / sum(B);
F1 = (0:7999)*Fs/16000;
h1 = freqz(B,A,F1,Fs);

for k = 1:length(freq_vec)

  freq = Fif + freq_vec(k);
  sig1 = 1000*exp(j*(2*pi*freq*t + 2*pi*rand(1)));

  sig2 = sig1.*exp(-j*2*pi*Fif*t);
  sig2 = filter(B,A,sig2);
  sig2 = sig2(500:end);  % Ignore transient

  ref = sig1(500:end);    % Ignore transient

% Phase estimation
  ph(k) = angle(sum(sig2.*conj(ref)));

end

% Group delay estimation
alg_grpdelay = -diff(unwrap(ph))*Fs/(1000*2*pi);

% Group delay estimation from freqz
gd = -diff(unwrap(angle(h1)))*Fs/(1000*2*pi);

% MATLAB group delay
gg = grpdelay(B,A,freq_vec,Fs);
```

N is chosen so that the time-bandwidth multiplication is above 1000, group delay response to the example in code 15.26 is as follows (Fig. 15.15):

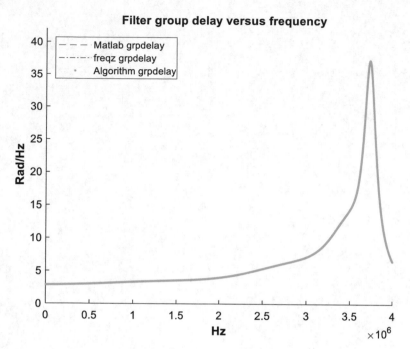

Fig. 15.15 Group delay estimation results

Implementation of algorithm elements such as complex exponent and NCO was described in Sect. 3.1.2 code 3.1.3, X.*conj(Y) in Sect. 12.2.1 code 12.5 and unwrap() in previous paragraph.

15.12 Conclusions

In this chapter we described the following problems in signal processing and how to solve them:

Dealing with input signal spikes

Dealing with input signal fast transitions

How to improve algorithm processing time

Center frequency estimation

Power level estimation

Noise floor estimation

Bandwidth estimation

Automatic level control

Frequency response estimation

Group delay estimation

Related Software

1. MATLAB version R2021a and Signal Processing Toolbox, The MathWorks Inc., Natick, Massachusetts, United States
   MATLAB and Simulink are registered trademarks of The MathWorks, Inc. See Mathworks.com/trademarks for a list of additional trademarks.
   For MATLAB and Simulink product information, please contact:
   The MathWorks, Inc.
   3 Apple Hill Drive
   Natick, MA, 01760-2098 USA
   Tel: 508-647-7000
   Fax: 508-647-7001
   E-mail: info@mathworks.com
   Web: https://www.mathworks.com
   How to buy: https://www.mathworks.com/store
   Find your local office: https://www.mathworks.com/company/worldwide
2. Microsoft Visual C++ 6.0, Copyright 1994–1998 Microsoft Corporation, Redmond, Washington, United States

© The Author(s), under exclusive license to Springer Nature Switzerland AG 2022
A. Dickman, *Verified Signal Processing Algorithms in MATLAB and C*,
https://doi.org/10.1007/978-3-030-93363-0

Index

Printed in the United States
by Baker & Taylor Publisher Services